Geographies of Mars

Geographies of Mars

Seeing and Knowing the Red Planet

K. MARIA D. LANE

The University of Chicago Press
Chicago and London

K. Maria D. Lane is assistant professor of geography at the University of New Mexico.

The University of Chicago Press, Chicago 60637
The University of Chicago Press, Ltd., London
© 2011 by The University of Chicago
All rights reserved. Published 2011
Printed in the United States of America

20 19 18 17 16 15 14 13 12 11   1 2 3 4 5

ISBN-13: 978-0-226-47078-8 (cloth)
ISBN-10: 0-226-47078-4 (cloth)

Library of Congress Cataloging-in-Publication Data
Lane, K. Maria D.
  Geographies of Mars : seeing and knowing the red planet / K. Maria D. Lane.
    p. cm.
  Includes bibliographical references and index.
  ISBN-13: 978-0-226-47078-8 (cloth : alk. paper)
  ISBN-10: 0-226-47078-4 (cloth : alk. paper)   1. Mars (Planet)—Research—
History—19th century. 2. Mars (Planet)—Research—History—20th century.
3. Mars (Planet)—Geography. 4. Mars (Planet)—Maps. 5. Martians.   I. Title.
  QB641.L366 2011
  523.43072—dc22   2010021769

♾ The paper used in this publication meets the minimum requirements of the American National Standard for Information Sciences—Permanence of Paper for Printed Library Materials, ANSI Z39.48-1992.

For Matt—
Everything, always.

CONTENTS

ILLUSTRATIONS

ACKNOWLEDGMENTS

Little did I realize how many debts could accumulate in the researching and writing of one small book. In looking back at those who contributed to this writing project, I am profoundly humbled to realize that my primary accomplishments are due largely to the generosity and talent of others.

First and foremost, I would like to acknowledge the impressive scholarship that precedes my own on the topic of Mars science and sensation. It is only through heavy reliance on works by Michael Crowe, Steven Dick, Bill Hoyt, Robert Markley, Bill Sheehan, and David Strauss that I have been able to grasp the complex history of Mars knowledge production well enough to explore some of its trajectories in further detail. I owe special thanks to Bill Sheehan for sharing his translation of Giovanni Schiaparelli's 1878 memoir, and to both him and David Strauss for their patience and enthusiasm in responding to questions, sharing ideas, and offering feedback on the plan for this book.

This project, first launched in 2000, has been funded at various critical junctures, making its continuation and eventual completion possible. Early archival research was funded by a Robert E. Veselka Endowed Fellowship provided by the Department of Geography and the Environment at the University of Texas at Austin and also by a George H. Mitchell Fellowship provided by the University Co-Op Society in Austin, Texas. The project's primary archival research phase was funded by an eighteen-month Mellon Fellowship for Dissertation Research in Original Sources provided by the Council on Library and Information Resources, with additional support from the American Historical Association in the form of a Beveridge Grant. The writing phases were supported by the Department of Geography and the Environment at the University of Texas at Austin and by a generous American Fellowship from the American Association of University Women.

This project also would have been impossible without the assistance of many extremely capable and accommodating professionals who hosted me in numerous libraries and archives. I am deeply grateful for the archival research assistance I received from Antoinette Beiser and Marty Hecht at the Lowell Observatory Archives, Brenda Corbin and Gregory Shelton at the U.S. Naval Observatory Library, Pam VanEe and Ed Redmond at the Geography and Map Division of the Library of Congress, Peter Hingley at the Royal Astronomical Society Library and Archives, Agnese Mandrino at l'Archivo Storico dell'Osservatorio Astronomico di Brera, and the especially helpful research staffs at the Cambridge University Archives, the British Library, the Boston Athenaeum, the Harvard University Archives, and the New York Public Library. Antoinette Beiser in Flagstaff and Agnese Mandrino in Milan, in particular, were extremely accommodating hosts, making significant contributions to this project that went well beyond their professional advice. I owe special thanks to Molly White of the Physics-Math-Astronomy Library at the University of Texas at Austin and to Donna Comer of the Centennial Science and Engineering Library at the University of New Mexico for helping me investigate numerous inquiries I would otherwise have given up as dead ends. Kris Doyle, Emily Doyle, and Molly Blumhoefer also provided invaluable research assistance. Richard McKim kindly provided reproductions of several important images, and various libraries granted permission to reproduce material held in their collections, as noted throughout the figure captions and notes.

The list of scholars and friends who have helped me at various stages in this project is surely longer than my memory allows, and I gratefully acknowledge that it rests on a broad base of substantial support. Some of the arguments included here appeared in early form in various publications and presentations, including a 2005 article in *Isis*, a 2006 article in *Imago Mundi*, and a chapter in the 2008 edited volume *High Places*. Material from the *Isis* and *Imago Mundi* articles (Lane, "Geographers of Mars," and Lane, "Mapping the Mars Canal Mania," respectively) appears in chapter 3, and parts of both chapters 4 and 5 may be somewhat recognizable as outgrowths of the *High Places* chapter (Lane, "Astronomers at Altitude").

For providing feedback on drafts of these publications and other early material, I thank Chris Gaffney, David Salisbury, Damon Scott, Maggie Lynch, Trushna Parekh, Lars Pomara, Jim Housefield, James Bryan, Steven Legg, David Lambert, Gerry Kearns, Matthew Edney, Denis Wood, Stephen Hanna, Wayne Prosser, Felix Driver, Bernie Lightman, Catherine Delano-Smith, Jason Dittmer, Veronica della Dora, Denis Cosgrove, and anonymous reviewers from both *Isis* and *Imago Mundi*. Denis Cosgrove, in particular, was an unexpectedly generous and enthusiastic supporter of this work, and I only wish that I could have finished the book before his untimely passing. For helpfully undertaking detailed readings of individual chapters, I am grateful to Karen Morin, Brad Cullen, Ian Manners, and especially Amy Mills, whose long-distance support helped me maintain a positive perspective on the sometimes difficult experiences of first-time authorship. For making the University of New Mexico an extremely pleasant place to work and write, I thank my colleagues Paul Matthews, Brad Cullen, Paul Zandbergen, Chris Duvall, Mindy Benson, John Carr, and Maya Elrick. And for providing critical and supportive mentorship throughout this project, I acknowledge Steve Hoelscher, Roger Hart, Diana Davis, and especially Ian Manners, who has given me the most splendid example to follow as a teacher and scholar.

The University of Chicago Press has certainly lived up to its reputation in all respects. For detailed comments that substantially shaped my thinking about this book and its arguments, I thank six anonymous reviewers recruited by the press. For patiently helping me through the publication process, I thank Therese Boyd, Stephanie Hlywak, Carol Saller, and Abby Collier. And for pointing me to UCP in the first place, I acknowledge a deep debt to Matthew Edney and Denis Wood, who practically gave me a testimonial about the many virtues of editor Christie Henry, none of which turns out to have been exaggerated.

Finally, I am grateful for support from the Lane and Doyle families over many years. Those closest to me have had more influence on this book than they probably realize, despite maintaining a healthy lack of interest in its specific claims. To Simon and Kristina, I can only express hope that we still have books and libraries when you are old enough to read this. And to Matt, I can only say thank you for making this book a part of your life, too.

ONE

Understanding Mars: Sensation, Science, and Geography

If future observations should confirm the views as to the artificial nature of these features of the surface of the planet which most nearly resembles our earth, it must be considered to be the most sensational astronomical discovery of the nineteenth century, and that which opens up the most exciting possibilities as to communication with beings who are sufficiently advanced to execute such widespread and gigantic irrigation works.

—British naturalist Alfred Russel Wallace (1898)

In an 1886 issue of the popular British magazine *Chamber's*, an anonymous article appeared under the title "Life on Mars." Declaring that Mars alone among the other planets of the solar system was likely to host life forms comparable to Earth's, the author suggested that Martian beings were probably fairly similar to humans. Despite the possibility of physical divergences conditioned by environmental factors, such as lungs capable of breathing extremely rarified air, the article asserted that there was no reason to think that the "Martialites" were anything other than thinking, sensing beings. In fact, the author reasoned, the considerably older age of the so-called red planet meant that "the Martialites are probably much further advanced in the arts and sciences" than humans. Recent reports from an "Italian astronomer who says he has lately detected lights on the planet moving about in such a way as seems to indicate a deliberate intention to open communication with the earth" were cited as support for this possibility.[1] Although much of the article was spent reviewing recent astronomical research and noting the many analogies that had been found between the geography of Mars and Earth, the author returned again and again to the question of Martian inhabitants and what they must be like. Reasoning that their form would be determined by the lesser gravity of Mars, the author offered the following comparison with humans:

> If, therefore, we assume that the men are of such a size that their weight and activity are the same as ours, they would be about fourteen feet high on the average. This would make their strength very great; for not only would it be actually superior to ours, but, as every weight is so much smaller, it would be apparently proportionally increased. We should, therefore, expect to find that the Martialites have executed large engineering works; perhaps also their telescopes are much superior to ours, and we have been objects of interest for their observers.[2]

Through this short assessment, *Chamber's* readers were treated to a preview of what would become the dominant interpretation of Mars over the following three decades. Although this particular unnamed author was likely drawing from a French astronomer's speculative interpretations of an Italian astronomer's recently reported Mars observations, these exact conjectures took root most deeply in the scientific and popular literatures of Britain and America over the next decade. The focus on Earth–Mars analogies, the speculation about Martian physical form, the certainty of Martian advancement in engineering, and the enthusiasm for Earth–Mars communication all figured prominently in the works of Anglophone astronomers, popular astronomy writers, literary commentators, and journalists alike. In what became a veritable sensation over Mars at the turn of the twentieth century, the physical and cultural geography of the red planet became household topics.

The beginnings of the fervor over Martian geography can be traced back to the 1878 publication of an Italian map that identified numerous linear features on the planet's surface. (See fig. 2.8 and the associated discussion in the next chapter.) No previous observer had detected anywhere near the level of detail that emerged on that particular map, and it would actually be another decade before other astronomers confirmed the straight, intersecting features it recorded in the landscape of Mars. Something about the exactness and linearity of the landscape that appeared on that map, however, captivated astronomers' attention while also provoking popular imaginations of Martian inhabitants. Initially labeled "canali" in the early Italian map, the faint lines came to be known as "canals" in the English-speaking world, soon anchoring a broader narrative that held the red planet to be both inhabited and irrigated by an advanced civilization. As this narrative became entrenched well beyond the confines of astronomy's disciplinary boundaries, it expanded to include the elements highlighted above in the 1886 *Chamber's* article. The Martians' apparent intelligence, size, and strength came to be seen as the means by which a vast network of irrigation canals had been engineered and built. The Martians' apparent organization and sophistication likewise spurred serious consideration of their ability and desire to communicate with humans.

Our modern view of the old "canal craze" holds that the original lines seen on Mars were probably an optical effect, in which astronomers' eyes resolved indistinctly seen landforms and color variations into simple shapes.[3] Today's satellites in Martian orbit and Rovers on the Martian surface have found no evidence of canals, vegetation, or advanced inhabitants. There is a tendency, therefore, to look back on the century-old maps with a kind of amusement, and to dismiss the conjectures about a super-race of canal-digging engineers as overly imaginative or even intentionally deceptive. Some of the astronomers involved have been painted as publicity-driven egomaniacs or theologically blinkered ideologues who positioned themselves in the debate according to personal agendas, regardless of the "evidence" they encountered in their observations and investigations of Mars.[4]

The most recent re-examinations of the historical record, however, have told a different story. As the topics of Martian geography were taken up by scientists, writers, commentators, lecturers, and artists a century ago, they were generally treated with seriousness and sometimes even deep philosophical attention.[5] As expressed in the epigraph that opened this chapter, for instance, one of the foremost naturalists and British public intellectuals of the time considered Mars-related science so "exciting" that it warranted inclusion in a book written to describe the nineteenth century's most important intellectual and technological developments. Although astronomers engaged in numerous debates as to what exactly they were seeing on Mars, how well they were seeing it, and how their unusual observations should be interpreted, the mere possibility of Martian inhabitants was so striking that it ignited significant reaction across other disciplines and audiences. Sometimes in agreement and sometimes at odds, these different individuals, institutions, and audiences contributed views that coalesced into a functionally dominant (if not universal) understanding of Martian geography as arid, inhabited, and irrigated during the two decades spanning the turn of the twentieth century. Although belief in Martian canals and in the possibility of intelligent life on Mars continued well into the twentieth century, there was a limited time in which Mars science engendered what can properly be labeled a sensation. By focusing on the nature of this limited popular phenomenon, this book aims to consider the larger process of scientific knowledge production that both informed and was constituted by the popular response to Mars reports.

Analogy and the Seeds of Sensation

The seeds of popular interest in Mars were actually laid very early in the scientific study of the red planet. Well before Italian astronomer Giovanni Schiaparelli reported seeing "canali" on the Martian surface in 1877, and well before American astronomer Percival Lowell interpreted the "canals" as evidence of intelligent life on the red planet in 1894, Mars was known to astronomers as essentially Earth-like. As intellectual historians have noted, the planet Mars was central to the Copernican revision of celestial mechanics that was confirmed by Kepler in 1607. In the consequent theological and philosophical consideration of other planets as potential "worlds" complete with intelligent inhabitants, Mars itself became one of the prime suspects.[6] In the ensuing debates over Mars' habitability and the probability of Martian beings' existence, commentators frequently turned attention to the red planet's perceived likeness to Earth, the only known inhabited planet. Although the early search for Earth–Mars analogies was driven by these philosophical considerations of Martian habitability, the potential for terrestrial analogy continued to drive scientific investigation of Mars for centuries (including into the present).[7]

Throughout the period of the turn-of-the-century Mars sensation that provides the focus of this book, many findings or claims about Mars were considered interesting primarily insofar as they contributed to or detracted from the long-assumed terrestrial analogy. The color of the dark patches on Mars was investigated in order to determine whether the planet had oceans or vegetation like Earth. The atmospheric thickness of Mars was investigated to determine whether there was sufficient air to support life such as existed on Earth. The seasonal variations in surface colors and patterns were investigated to assess whether Mars experienced cycles of vegetative growth and senescence similar to those on Earth. All of these investigations were ultimately based on the original question of whether Mars was habitable or inhabited. As American astronomer Edward Holden put it, "There is certainly no more important question in planetary astronomy than to determine whether our neighboring planets are or are not inhabited. . . . To solve this question it is necessary to construct the most accurate map of the planet's surface and to observe with the greatest care all the phenomena as well as possible by means of terrestrial analogies, if this be possible."[8] In essence, analogy became a fundamental way of thinking about Mars rather than merely a way of describing it. By the late nineteenth century, Mars was typically referred to as Earth's "nearest neighbor" or the planet in the solar system with "the greatest analogy" to Earth, despite the fact that Venus was commonly known to be closer to Earth in both size and orbital distance.

To a large extent, this phrasing reflected a focus on visible landscapes, the category in which Mars was clearly thought to be more analogous and interesting than the cloud-enshrouded Venus:

> Though little more than half the Earth's size Mars has a significance in the public eye which places it first in importance among the planets. It is our nearest neighbor on the outer side of the Earth's path round the Sun, and viewed through a telescope of good magnifying power shows surface markings suggestive, with the aid of imagination, of continents, mountains, and valleys; of oceans, capes, and bays, and all the varying phenomena which the mind readily associates with a world like our own.[9]

Descriptions and interpretations alike relied on such visual analogy, casting Mars as a landscape that could be observed in the same way travelers and geographers examined Earth's visible landscapes. The strength of the general Earth–Mars analogy was thus bolstered by representations of Martian landscape and culture as explicitly similar to exact locations and peoples on Earth. Sir Norman Lockyer, an eminent English astronomer, described sketches of Mars thus in an astronomy textbook: "In the upper [drawing] a sea is seen on the left, stretching down northwards; while, joined on to it, as the Mediterranean is joined on to the Atlantic, is a long narrow sea, which widens at its termination. . . . The coast-line on the right strangely reminds one of the Scandinavian peninsula, and the included Baltic Sea."[10] Lowell likewise compared the size and probable operation of the Martian canals to the well-known waterway at Suez and contrasted their geometric appearance with the winding Mississippi River. He also frequently used terrestrial metaphors for literary effect, as when he remarked that a feature appeared to be "a beautiful cobalt blue, like some Martian grotto of Capri."[11] Many other Mars observers equaled him in this regard, with various Martian features compared at one time or another to Switzerland, Ireland, Amsterdam, London's Hyde Park, Ohio, Puerto Rico, Scandinavia, the Mediterranean Sea, the Strait of Malacca, Lake Tanganyka, the South African veldt, and so on. Such comparisons generally served to "tighten the knot of analogy between Mars and the Earth" and reinforce the idea that Mars was "a small version of the Earth."[12]

Even when claiming that Mars was rather different from Earth, astronomers typically reinforced their arguments with analogies to specific places. For instance, Schiaparelli wrote in 1893 that the general topography of Mars "does not present any analogy with the Earth" but then continued that the canals could be "produced by the evolution of the planet, just as on the Earth we have the English Channel and the Channel of Mozambique."[13] Similarly, Holden argued in a critique of Lowell that terrestrial analogies failed to explain the changes on Mars, but then in the same paragraph suggested a terrestrial analogy to explain the faintly colored regions of Mars: "Are they vast shoals like the Grand Banks of Newfoundland?"[14] The repeated invocation of specific terrestrial landscapes thus paradoxically served mainly to reinforce the widespread conviction that Mars could be explained almost entirely by analogy with Earth.

These comparisons served in general to support the emergence and duration of the sensational inhabited-Mars theory. Upon reading that "the smallest object that would be discernible on Mars must be as large as London [and that] it would not be possible to see a point so small as would either Liverpool or Manchester be if they were on that planet," readers had to make only the smallest conceptual leap to imagine actual Martian cities.[15] Similarly, reports that the annual melting of Mars' polar ice caps "is of as much importance as the annual inundation of the Nile is to the Fellaheen of Egypt" helped cast Mars as a specific, legible, populated landscape.[16] Lowell, in particular, used the Mars–Earth analogy eloquently in support of his arguments, inspiring readers' interest in the possibility that Mars could be an inhabited world:

> For all practical purposes Mars is our nearest neighbor in space. Of all the orbs about us, therefore, he holds out most promise of response to that question which man instinctively makes as he gazes up at the stars: What goes on upon all those distant globes? Are they worlds, or are they mere masses of matter? Are physical forces alone at work there, or has evolution begotten something more complex, something not unakin to what we know on Earth as life? It is in this that lies the peculiar interest of Mars.[17]

Markley has argued that the Earth–Mars analogy operated paradoxically, working at the scale of the planetary whole yet breaking down at the scale of specific landforms like the canals.[18] Even acknowledgments that the Mars–Earth analogy was imperfect, however, do not seem to have dimmed the overall enthusiasm for terrestrial comparisons or for Martian habitability. Referring to the work of several astronomers who disputed the inhabited-

Mars theory and its analogical basis, for example, the Welsh astronomy writer Arthur Mee admitted, "on the whole, their testimony does not make in favour of terrestrial analogies, which seem to diminish, the closer and more critical the examination of the planet." At the same time, however, Mee wrote as if convinced that the analogy was correct: "the general aspect of the planet reminds one strangely of the probable appearance of our earth could we view it at the distance of Mars. On the rare occasions when I have been fortunate enough to secure good views of the planet, the impression of sea and land and polar snow was overwhelming."[19]

Crowe has asserted that logical fallacies—such as the mistaking of analogy for proof—were instrumental to most of the claims made by early proponents of the inhabited-Mars hypothesis.[20] But visual analogy was much more important than merely providing a plausible substitute for logical proof. It produced scientific understandings and provoked popular sensations that gave Mars a specific cultural significance that would not have developed otherwise. Although viewers like Mee could concede that Mars' geometric surface features defied analogical explanation, they still maintained that the general appearance and seasonal variations of Mars indicated a living world that could host intelligent life. It was this paradox that encapsulated the intrigue Mars presented to popular audiences. If Mars was fundamentally similar to Earth, yet also radically different, the challenge of making sense of its landscape became both daunting and imperative.

The Popularization of Martian Geography

Building on fundamental and long-standing assumptions about the many analogies between Earth and Mars, the "sensation" over Mars really began in the 1890s. In the spring of 1895, American astronomer Percival Lowell began publishing for the first time on Mars research he had initiated the previous summer at his brand-new Flagstaff observatory. In tandem with a blizzard of articles that began to appear in periodicals like *Atlantic Monthly*, *Scientific American*, *Popular Astronomy*, and *Astronomy and Astro-physics*, Lowell also targeted broad audiences with a wildly successful lecture tour on the East Coast. In his presentations Lowell explained both the science and significance of the apparent canals on Mars, which he interpreted as irrefutable evidence of intelligent life. As the *Boston Commonwealth* reported, the events held in Boston drew "crowded audiences of people who filled every seat and all the standing room in Huntington Hall."[21] These crowds reportedly went away thoroughly convinced by Lowell's arguments, largely because he astutely presented the material in a way that inspired confidence in his science while yet sparing his lay audiences the dry technical presentation they might expect on a scientific topic.

> It is impossible in print to describe the charm of Mr. Lowell's lectures. His humor, his ready wit, his complete knowledge of the subject with which he deals, are such as one has no right to expect in the same public speaker. The most serious considerations are made interesting by analogies with affairs with which we are familiar and in which we are at ease. Everybody knows how light his pen is when he writes of his travels, and his ease as a public speaker and the readiness with which he takes his audiences into his confidence give an additional charm to the lectures as he reads, or rather, as he delivers them.[22]

In this accessible and analogy-filled speaking style for which he became known, Lowell provided an overview of scientific concerns and techniques for studying Mars but proceeded quickly to a broader interpretation of his results, offering a somewhat metaphysical view of his predicted Martian inhabitants.

> Mr. Lowell, with great humor but with absolutely accurate mathematics, showed to his hearers how large and tall and strong the Martian people might be. The attraction of gravitation is only one-third what we have here. . . . The physical power of this man is as great in proportion, his memory of the past may be more accurate, as it would seem that his foresight for the future is more sweeping. So it is that a population quite as dense, we may believe, as the population of this world, a population which has not spent, apparently, most of its history in mutual throat cutting and constant quarreling, has achieved the marvels of irrigation and vegetation which we see upon the planet Mars to-day.[23]

As this enthusiastic report indicates, popular audiences typically gravitated toward the most speculative and interpretive elements of Lowell's presentation, finding cultural significance in his conjectures about Martian bodies and civilization.

His willingness to discuss the implications of Mars science, rather than limiting himself to a discussion of proven (or even prove-able) facts, however, strongly tainted Lowell's relationship with the community of professional astronomers. Leading American astronomers complained to one another in private correspondence about his activities and influence, sometimes even publishing responses in the popular press that were meant to caution Lowell's audiences about the dangers of jumping to conclusions about Mars.[24] In the process, they hoped to protect the scientific reputation of their discipline by exposing Lowell as an amateur whose claims were unsound and unscientific, based on little more than speculation and optical illusion. Despite increasingly hostile accusations from the professional ranks, however, Lowell maintained a very high profile with nonspecialist audiences. He was embraced by the popular astronomy journals and made no apology for his efforts to establish Mars as a topic of philosophical and sociological importance.[25] In fact, his observatory records show that Lowell hired a publicist for some period of time to help him reach wider audiences and direct public opinion in favor of his theories.[26] These activities were bolstered by an oft-cited defiance toward his critics, whom he chastised for an inability to communicate with general audiences, noting caustically that "it is [not] so hard to make any well-grasped matter comprehensible to a man of good general intelligence."[27]

Although numerous attempts to discredit Lowell within the discipline of astronomy had finally begun to take their toll by 1909, his influence on popular audiences did not diminish nearly as quickly as his reputation in the disciplinary power centers. Lowell continued to lecture on Mars and other topics and was reportedly even pressed to run for political office in Arizona.[28] Upon his death in 1916 Lowell was eulogized as "the man who made the Martians live," who "gave untrained men the freedom of the skies, turned the imagination of nations starward and enlarged our conception of what the life of the universe may be."[29]

Lowell's experience illustrates the complex ways in which scientific popularization allows a coherent body of knowledge to emerge from the work of individuals and audiences who consider themselves mutual antagonists. First and foremost, Lowell reminds us that popularization must be seen as a fundamental part of the process of knowledge production. Although traditional views of science popularization long posited a distinction between "genuine" scientific knowledge and "distorted" popularized knowledge, some now-classic investigations of this distinction have shown that there are actually no such boundaries that can be delineated.[30] Scientists often learn about others' work (and about reactions to their own) from popular sources, "and these shape their beliefs about both the content and the conduct of science."[31] Scientists are thus influenced by popular audiences' consumption of scientific claims just as surely as popular audiences are swayed by the apparent irrefutability of scientists' statements. The "popularizers" themselves—those who restate technical arguments in more accessible and often more interpretive ways—thus play an important role in brokering the production of specific knowledge claims. Lightman and others have noted that popularizers became particularly important at the end of the nineteenth century precisely because many sciences had by that time begun to differentiate "professional" scientific work from "amateur" involvement. As the disciplines accordingly became more specialized, "the need arose for nonprofessionals, who could convey the broader significance of many new discoveries to a rapidly growing Victorian reading public."[32] In the process of restating scientists' ideas and claims, however, popularizers gained some measure of control over them. As a result, scientists soon found themselves forced to confront what they sometimes considered misinterpretations of their work or improper emphases on unimportant elements of their research. In the process, time and attention could be diverted from original research agendas. In this way, the production of scientific knowledge was (and is) thoroughly bound up with popular consumptions of and responses to this knowledge.

In the case of Mars, a number of popular writers took the stage before Lowell. French astronomy writer Camille Flammarion dominated popular understandings of Mars in the mid- to late nineteenth century, compiling in 1892 a comprehensive chronology of all known Mars investigations since the 1600s. In presenting the history of Mars observation, Flammarion took a firm stance in ongoing arguments about humanity's uniqueness in the universe, arguing that Mars was very likely to be inhabited. Richard Proctor, an English astronomy writer, also focused on the probability of Martian inhabitants, concentrating particularly on the efforts of English astronomers such as William Dawes.[33] Both Flammarion and Proctor had conducted their own celestial and planetary observations. They built their reputations, however, around their ability to present astronomical advances in language and formats accessible to broader audiences. But as scientific professionalization and astronomical specialization took place in the late nineteenth century, it became increasingly difficult to assume the roles of both astronomer and popular writer. Perhaps for this reason, Lowell's explicit and unapologetic participation as *both* a producer of scientific knowledge and an interpreter of this knowledge was seen as more problematic than it had been for those who went before him. At the same time, Lowell also managed to inspire the most emphatic and enduring popular fascination with the red planet in a century.

As the Lowellian prospect of an inhabited Mars blossomed into a popular sensation—discussed in newspapers, lectures, books, plays, poems, cartoons, and other broad-circulation media—incredulous astronomers found themselves drawn into Lowell's orbit against their wishes. This sensation influenced professional scientists in two main ways. First, it forced leading astronomers to respond to what they considered outrageous and speculative commentaries that had begun to circulate in mainstream publications. Lick Observatory director Holden, for example, wrote an 1895 article in the *North American Review* explicitly meant to dispel the popular sensation, saying:

> It is very unsatisfying, no doubt, not to be able to answer many questions which an eager mind can put regarding the planets, their constitution, their habitability, etc. But it is most satisfactory to have this year taken the very important step of clearing the way for such solutions by sweeping out of sight the fabric of assumption and ungrounded assertions which has lately barred the path.[34]

His colleague, James Keeler, at the Allegheny Observatory likewise acknowledged in *Century Magazine* that speculative writing could be amusing, but cautioned that "it is well to remember the slightness of the foundation on which our superstructure is reared."[35]

Many of these astronomers clearly wished that all the time and energy invested in debating Mars could be diverted to other, more meaningful topics. Director of the American Nautical Almanac Office Simon Newcomb, for instance, chastised Percival Lowell in personal correspondence for focusing so much attention on Mars when he could have been making other astronomical advances:

> Are you not well situated for making better observations of the spectra of zodiacal light than any heretofore obtained? Another class of desirable observations is the most exact observations practicable of the position of the axis of the light through a period of an entire year. It is most astonishing to me that when people are making so many sporadic observations, which are of no importance whatever, all the stations near enough to the equator to make a continuous series of value are doing nothing in the matter.[36]

Thanks to the refusal of Lowell and his supporters to heed these admonitions, however, professional astronomers grew increasingly frustrated in their attempts to control the sensation. As one amateur astronomer boldly wrote in *Knowledge* in 1901: "To refrain altogether from speculative hypotheses would be as unscientific as uninteresting; the sensational theories about Mars have been a stimulus to much excellent work."[37] With Lowell and a sympathetic public in this mood, the sensation would not fade quickly.

A second effect of popular interest on professional astronomical work was that it opened a suite of new publications and venues to the circulation of scientific claims about Mars. Lowell himself was a master at targeting popular audiences through literary magazines, lectures, and even curated exhibitions. His 1907 display of Mars photographs on the Boston campus of MIT drew enough of a crowd that a Harvard staff astronomer was moved to note Lowell's "special effort . . . to keep the public informed," even as he cautioned that the photographs themselves were inconclusive in terms of proving or disproving the existence of canals on Mars.[38]

Despite concerns about the distorting influence of Lowell's preferred venues and formats, however, many professional astronomers found themselves reluctantly offering their own articles to the mainstream periodicals. In the wake of Lowell's high-profile expedition to the Andes Mountains in 1907 to capture many of the photographs later displayed in Boston (discussed in chapter 4), for instance, astronomers William Pickering and Simon Newcomb both contacted *Harper's* separately with articles they hoped the magazine would publish.[39] By taking their refutations of Lowell directly to one of the most dominant general-interest periodicals then in American circulation, these astronomers were clearly influenced by the popular sensation Lowell had generated. It also reached into the pages of the major astronomical journals themselves, as editors began to reprint popular articles for their specialized audiences. As an example, one 1895 article about Mars science appeared in three different publications over a few short months. Hale's "Latest News from Mars" appeared in *Scientific American*, then was reprinted in the *Boston Commonwealth*, and finally appeared again as a reprint in *Publications of the Astronomical Society of the Pacific*. Although these reprints were often intended to ridicule or lament the dangers of sensationalism, the publication of and response to popularizers' work in these formal disciplinary venues make clear that ideas were circulating in both directions.

As popular interest grew and provoked increasingly speculative or bizarre commentary in the mainstream press, therefore, it became more and more influential on the production of scientific knowledge itself. The popular sensation over Mars as an inhabited planet thus emerged from and remained linked with a fairly technical scientific narrative for two decades. Even as popular audiences' interests ran far afield of specialist astronomers' interests, the two strongly influenced one another in multiple ways. They should not be viewed as distinct and separate types of knowledge, but rather as mutually constituted competitors in a comprehensive knowledge production process.

Understanding Martian Geography

Recent discussions of current Mars science have acknowledged that the late-nineteenth century treatments of Mars—in both science and popular culture—continue to influence the ways we study and understand Mars today.[40] This book represents an attempt to make sense of this knowledge production process during the decisive time period in which the red planet was widely thought to be arid, irrigated, and inhabited. What exactly did scientists claim about specific Martian landscape features during this time period? How exactly did they cultivate legitimacy for these claims? How did they mobilize support from readers, audiences, and scientific allies? How, in turn, did the need for support and the reactions of potential supporters influence the nature of the claims that could be made about Martian physical and cultural geography? And what does any of this have to do with Mars itself, if anything?

In the end, this book is not really about Mars or even about Martian geography. It is about the processes through which geographical knowledge was (and is) produced in specific places at specific times by specific individuals, institutions, and publics. In exploring the intellectual, social, and geopolitical contexts that gave rise to a widespread belief in canal-digging Martians around the turn of the twentieth century, I am more essentially concerned with the nature of those specific contexts themselves than with any specific landscapes of Mars. The argument throughout this book holds that the geopolitical moment in which the inhabited-Mars narrative unfolded—dominated as it was by European imperialism and American expansionism—produced an intellectual and social climate in which the view of Mars as an arid, dying, irrigated world peopled by unfathomably advanced beings was really the *only* interpretation of Mars observations that could plausibly have been accepted by large numbers of Western scientists, writers, and audiences. This conclusion is reached largely by pursuing a geographical approach that considers the spatial dimensions of Mars science itself, as well as the ways in which specific claims about Mars intersected with emergent trends in the discipline of geography. To that end, the book focuses almost exclusively on British and American scientists and audiences to more fully explore how the scientific debates over inhabited-Mars theory were enmeshed within a geopolitical context in which Britain and America were key players. Some key continental scientists are included (primarily Italian astronomer Giovanni Schiaparelli and French astronomer/writer Camille Flammarion), but only in the limited context of their influence on British and American audiences through citations, publications, correspondence, and news reports.

In focusing a tight geographical lens on what is already a fairly notorious and well-studied episode in the history of science, this book inevitably skims lightly over several important influences that other scholars have already highlighted. The roles played by social-institutional dynamics within disciplinary astronomy, by psychological-perceptual factors, and by ongoing debates over theology and evolution, for example, are explored primarily in terms of their relationship to the geographic contexts that are foregrounded here.[41] Perhaps most grievously, this focus privileges geography over other nonastronomical sciences (especially geology and ecology) that also impacted the practice and reception of Mars science during a period in which the natural sciences had not yet fully formed their distinct identities. The success of recent works examining the development of biological and ecological metaphors in Mars science, however, suggests that a similarly productive result awaits the careful study of Mars from a geographical point of view.[42] As I hope the book shows, this disciplinary focus reveals that the functions and meanings of Mars science were deeply influenced by an engagement with geographical practices and ideas.

Mars and the Geography of Science

In studying the Mars sensation as an episode of knowledge production, I draw theoretically from critical scholarship in the history of science, which has now convincingly demonstrated that there is "nothing self-evident or inevitable" about scientific claims that become established as "truths" in specific times and places.[43] Early constructivist critiques of science argued that scientific knowledge should be understood "primarily as a human product, made with locally situated cultural and material resources, rather than . . . simply the revelation of a pre-given order of nature."[44] This scholarship, which used sociological methods to show that scientific knowledge claims were more closely related to social contexts than to any objective "reality," usefully focused attention on the relationship between knowledge and power. At the same time, however, it also often tended to obscure the ways that individuals engage with real phenomena in unique places and reach conclusions with some hesitation.[45] In response to this shortcoming, critical scholarship in the history of science has now gone beyond constructivist assertions regarding the cultural dimensions of scientific knowledge, formulating instead a model of science *as* culture. In rereading and revising traditional historical accounts that treat science and culture as separate entities, this recent approach to the study of science suggests that scientific change occurs as a result of complex cultural negotiation. Identifying numerous instances of "translation" and "hybridization" (subversive appropriation) of one culture's knowledge/power claims by another, recent scholarship thus avoids both universalism and relativism by focusing on the complex social, political, and religious positioning of various actors engaged in the translation or modification of cultural-scientific knowledge claims within specific historical contexts.[46]

In one of many strands in this scholarly development, recent work has turned considerable attention to the "geography of science" as a contextual dimension that has been largely ignored in scholarly understandings of how scientific truths are formed. The classic works in science studies, which focused on the contingent and situated nature of scientific knowledge, essentially begged a geographical consideration without addressing it directly. Early work in science studies showed that the emergence and institutionalization of experimental science, for example, was dependent on the gathering of "witnesses" who could vouch for the legitimacy of experimental observations and phenomena. The uniquely local laboratory sites in which witnesses were typically assembled are now understood to have reflected and replicated social geographies of privilege. This spatial expression of a social geography thus allowed for the cultivation of "trust" in the truth of scientific claims, even among those who had not witnessed the reported empirical phenomena in person.[47]

Despite these early acknowledgments of spatial influences in the practice of science, the "geographical turn" in this literature is just now coming into full swing. Ophir and Shapin helpfully suggested that the "irremediably local dimension" of scientific knowledge should be seen not as a damaging critique but as a methodological point of entry.[48] Livingstone has also pushed for an explicitly geographical approach to the study of science, inspiring a substantial body of scholarship to this end. Arguing that "[s]cientific notions like discovery, the challenge to authority, natural knowledge and so on both *produce* and *are produced by* geography," Livingstone has called for attention to

the role of the spatial setting in the production of experimental knowledge, the significance of the uneven distribution of scientific information, the diffusion tracks along which scientific ideas and their associated instrumental gadgetry migrate, the management of laboratory space, the power relations exhibited in the transmission of scientific lore from specialist space to public place, the political geography and social topography of scientific subcultures, and the institutionalisation and policing of the sites in which the reproduction of scientific cultures is effected.[49]

The places in which scientists conduct their work, the pathways and networks along which scientific claims travel, and the unique locations in which audiences engage with scientific knowledge have all now been shown to influence not only the substance of scientists' work, but also their ability to gain credibility.[50] The spatial settings in which scientific work is undertaken are thus no longer viewed "as passive backdrops, but as vital links in the chain of production, validation and dissemination."[51] This scholarship helpfully nudges us beyond the problematic constructivist-versus-realist debate over the "true" nature of "science" by acknowledging the plural and varied natures of science, scientists, scientific investigation, and scientific knowledge claims in their many spatial variations.

Livingstone's repeated calls for attention to science's geography have focused on three major themes: site, region, and circulation.[52] The sites in which scientists work are now acknowledged to have a fundamental impact on the way knowledge claims are constructed and prepared for dissemination. Important recent work on this theme has shown that scientific knowledge is produced in a multiplicity of sites, including not only the controlled laboratory, but also, for example, the field, museum, hospital, pub, coffeehouse, bazaar, ship, and body. The microgeographies affecting each site of science are "central to the veracity of the knowledge produced," despite the common perception of science as a "placeless" activity that does not vary by location.[53] At a broader scale, regional geographies influence not only how scientists will approach their work, but also how that work will be received. The role of local scientific societies, for instance, has important regionally specific effects on the legitimization of scientific work.[54] Finally, geographies of circulation among sites, regions, and audiences are now seen as important determinants of scientific knowledge and its credibility. Because scientific practices are typically separated from witnesses or audiences by some spatial distance, the establishment of trust (and, therefore, legitimacy) usually requires a circulation of knowledge claims. The spatial- and social-geographic dimensions of this circulation influence the nature of the claims themselves, as well as their reception.[55]

In revisiting the turn-of-the-century process of knowledge production for Mars, it is clear that a geographical perspective informed by this scholarship could do much to expand our understanding of how and why certain claims about Martian geography gained such widespread footing. Astronomers and observatories moved from place to place in search of better observing conditions. Antagonistic observers tried to discredit one another based on the locations of their observatories. Mars-observing expeditions sent to tropical mountains generated sensational newspaper coverage almost independently of their claims about Mars. And audiences in different locations responded with markedly different levels of enthusiasm to the inhabited-Mars theory. Although some of these spatial dimensions have been noted in previous scholarship on the Mars canals, none has been addressed explicitly from a geographical point of view. Many of the intended contributions of this book can therefore be attributed to a simple shift in perspective that puts place and spatial relationships into a central analytical category that takes priority over any emphases on personality, philosophy, theology, rhetoric, or objectivity.

Mars and the Science of Geography

As traced in the chapters that follow, a focus on the geography of Mars science leads to a parallel focus on the science of geography itself. Knowledge about the canals was first circulated in maps—the quintessential geographical representational format—that strongly influenced the nature of both scientists' and audiences' reception of the inhabited-Mars theory. Astronomers' claims about Martian surface characteristics were likewise often anchored by the latest geographical theories, and popular audiences encountered news about Mars in the same periodicals (sometimes even on the same pages) that carried news from terrestrial geographic expeditions. Additionally, quite a few of the more prominent Mars astronomers were actually associated with geographical work and participated in social networks that included geographers. The Italian astronomer Schiaparelli, who published the first canal-map of Mars, for instance, published on the meteorology and topography of Milan, and his personal papers show that he corresponded extensively with Italian and other European geographers.[56] Similarly, the director of the American Nautical Almanac Office, Simon Newcomb, who became heavily involved in the Mars debates as a proponent of the optical illusion theory, corresponded with American geographers and even served as an adviser to President Theodore Roosevelt on a proposed expedition to the Philippines.[57] Percival Lowell, who was responsible for the broadest dissemination of Mars-related knowledge to the American and British publics, enjoyed a decade-long career as a travel writer and sometime-diplomat in East Asia before he turned his attention to astronomy and Mars.[58]

In looking at how a geography-of-science approach might be applied to Mars-science-as-geography, however, there are few examples to follow.

Although geographical themes have now become commonplace in the science studies literature, historical geographers have just begun to apply them to their own discipline.[59] To a large extent, these applications have been pursued in a vein of "critical" history that has struggled to explore the production of geographical knowledge as "a situated social practice." Following a turn toward "contextual" histories of geography that were more mindful of the discipline's imperial connections than the traditional great-man narratives, the more recent push toward critical history has relied heavily on social theory to explore "the various ways in which geographical knowledge has been implicated in relationships of power."[60] The most obvious focus of inquiry for many of these scholars has been the foundational relationship between geography and imperialism. From "territorial acquisition, economic exploitation, militarism and the practice of class and race domination," the geopolitical interests of European nations led directly to state investments in academic departments of geography, which became institutionalized during the period of most aggressive imperial activity in the late nineteenth century.[61] In return, the mandates of imperialism had a significant ideological influence on the newly established discipline, as the most widely held theoretical approach to geography during this period of institutionalization—environmental determinism—was grounded in a racist ideology used to justify the colonial administration of tropical realms by white imperialists.[62] Although historical geographers and historians of geography alike have attempted to explore these relationships more fully, they have often been constrained by a focus on discourse and text to the exclusion of tangible, material geographies.[63] Environmental historians have been more successful in this regard, linking geographical discourses and institutions more concretely with the transformations of colonial landscapes.[64] Building on this scholarship, more overtly material and spatial approaches from within historical geography itself have also now begun to address the challenge of writing geography's history from a more critical point of view.[65]

By tracing the geography of Mars science—through the movements of its practitioners, the sites of its knowledge production, the circulation of its claims, and the locations of its variously enthusiastic or skeptical audiences—these pages lead to the inescapable conclusion that disciplinary geography and its imperial influences were fundamental to the emergence, entrenchment, and duration of the inhabited-Mars hypothesis. There were, of course, other important intellectual and institutional contexts that also influenced the production of Mars-related knowledge, but key developments in both the science and sensation of Mars often turned on issues of geographic thought and method. Re-examination of the historical record covering the decades of Mars sensation thus provides additional insight into the nature of geographic knowledge and the discipline of geography itself. During a time of major cultural, intellectual, political, and economic transition in the Western world, when European imperialism was at its zenith and American expansionism had begun in earnest, the construction of Mars as an incomprehensibly complex and engineered world both reflected and challenged some of the most dominant themes in the discursive repertoire of imperialism. Closer examination of these themes from a geographical perspective thus promises a new way of thinking about geography's spatial dimensions during the imperial age.

The Rest of the Volume

The following chapters trace the intellectual, practical, and representational reverberations between geography and astronomy as they emerged in the production of Mars-related knowledge. Chapter 2 emphasizes and explores the role cartographic representation played in establishing early views of the Martian landscape as irrigated and inhabited. Starting in the mid-nineteenth century, most astronomers recorded and disseminated data about Mars in cartographic formats. Not only did this practice contribute to the perception that Mars science was a geographical endeavor on par with numerous imperial mapping projects, but it also produced a particularly authoritative view of the red planet as an inhabited landscape. Relying primarily on archival and published texts, images, and maps, chapter 2 reviews the conventions and transitions of Mars mapping from 1877 to 1910. It explains how linear details first appeared on astronomers' maps, why they persisted as authoritative representations of the Martian landscape, and the ways in which they induced astronomers to accept the inhabited-Mars hypothesis. The chapter's analysis of maps, cartographic icons, and the textual narratives that supported them exposes the extraordinarily powerful role cartographic representation played in turn-of-the-century Mars science. This power is further underscored in the concluding discussion, which reveals that challenges to the authority of astronomical cartography—primarily from new photographic techniques developed in the early 1900s—exactly coincided with the waning of the Mars canal sensation. Cartography is thus shown to be a critical foundation of legitimacy for the inhabited-Mars hypothesis that also linked to nationalistic and imperial competitions.

Chapter 3 focuses on the locations in which Mars science was conducted. The Mars debates unfolded during a critical period in astronomy's history marked by the building of new observatories, the professionalization of observatory staff, the emergence of expert specializations, and the transition from planetary to stellar topics. As this transition developed, high-altitude and mountain sites became a locus of legitimacy in North American and European astronomy, particularly favoring new observatories in the American West. The leading voices in the scientific Mars debate, in fact, emerged from these new observatories and relied heavily on representations of their mountain locations to establish credibility for their claims about Mars. The analysis presented in this chapter uses both published and archival sources to explore the nature and impact of these representations. It traces the development of rhetoric and imagery related to high-altitude vision and work in publications about Mars, showing how such representations resonated with both scientific and general audiences. It also discusses the ways that astronomers working on Mars topics disrupted their discipline's attempt to set universal legitimacy standards. By casting their legitimacy in terms of specific sites and personal qualities, Mars astronomers essentially removed the question of replicability from evaluation of their scientific claims. This development explains both why the debates over Mars caused serious consternation among the emerging class of professional astronomers and why professional astronomers had such difficulty combating methods and claims they considered unscientific. In these debates, mountain representation played a critical and complex role in the cultivation of legitimacy.

Chapter 4 shows that the most influential Mars astronomers were often represented as heroes, travelers, and adventurers. In the same way that astronomers working in mountain sites became more authoritative than their sea-level counterparts, astronomers who successfully presented themselves as explorers or geographical expeditioners managed to control the Mars debates most deftly. This chapter explores several elements of this phenomenon, using both published and archival materials to follow the activities and representational practices of three different Mars-related expeditions conducted by American astronomers. It shows that astronomers' use of a geographical persona to legitimate their claims allowed them to rely on audiences' strong preconceptions about geographical exploration and heroic adventuring. Furthermore, it actually influenced the ways astronomers thought about and conducted their scientific work. The chapter concludes that astronomers' choices about how to represent themselves largely determined their success in convincing professional colleagues and popular audiences that their claims about Martian geography were legitimate.

Chapter 5 moves from a focus on scientific practice and representation to a more sustained examination of the cultural and intellectual meanings

Mars science carried at the turn of the century. Using points of debate to crystallize the analysis, the chapter is structured around the contrasting viewpoints of two well-known public intellectuals—one British and one American—who participated vigorously in the interpretation of Mars science. As Percival Lowell and Alfred Russel Wallace argued about how to interpret the most recent Mars science, they relied heavily on geographical and political theories that were then emerging as the drivers of imperial and expansionist policy and activity. From landscape change to cultural hierarchy to irrigation, the most widely remarked themes stemming from the inhabited-Mars hypothesis had strong connections to contemporary debates regarding the control of terrestrial landscapes and peoples.

Chapter 6 draws from the insights of the previous chapters to argue that the popular sensation over Martian landscapes and inhabitants constituted a meaningful engagement with imperial and expansionist imaginative geographies. Highlighting the differing responses of British and American audiences to the most speculative elements of the inhabited-Mars hypothesis, it suggests that fundamentally divergent approaches to changing geopolitical environments can be read through the most prominent scientific and popular narratives about Mars. Analyzing published materials from scientific sources, news media, and both highbrow and mainstream fiction, this chapter uses the lens of national difference to explore the multiplicity of imaginative geographies that can be read in the phenomenon of turn-of-the-century Mars science.

TWO

Representing Scientific Data: Cartographic Inscription and Visual Authority

This whole wonderful map produces the absolute and irresistible conviction, that these "canals" owe their existence to a guiding intelligence.

—American inventor Nikola Tesla (1907)

At the root of the inhabited-Mars narratives lay a series of detailed maps. In addition to serving as graphical data repositories, these maps served several complex functions in the development of Mars' scientific and cultural meanings. Maps solidified the idea that Mars was like Earth, induced nationalistic competitions among astronomers, and authorized a view of the Martian landscape as modified and possibly inhabited. In doing so, Mars cartography profoundly influenced the nature of planetary investigation and contributed to an unprecedented scientific and popular acceptance of the possibility that life might exist on worlds beyond Earth.

This chapter identifies a series of transitions in generally accepted cartographic conventions that bracketed the popular Mars sensation and played a central role in the debates over Martian geography. Close examination of a mapping controversy in 1877–78 shows how established understandings of Martian geography were challenged by new cartographic views that laid the foundations for the inhabited-Mars hypothesis. Competitions to decide which of two high-profile maps was "right" unfolded in a complex manner, influenced by trends in other sciences and in cartography more broadly. Although no single standard map of Mars ever emerged, the post-1877 dominance of a specific style of mapping contributed critically to both scientific and popular belief in intelligent Martian inhabitants. Not until 1909, when new forms of astronomical photography began to challenge the authority of maps, did the most sensational era of belief in Martian inhabitants come to an end.

The most popular explanation of the long-running canal craze holds that the whole episode rested on a simple mistranslation of the word "canale" from original Italian maps. If English translators had eschewed the artificial-sounding word "canal" and chosen instead the more appropriate and natural-sounding word "channel," the argument goes, it is unlikely that so many people would have developed a mistaken impression about life on Mars. American astronomer Simon Newcomb went so far as to say at the time that the "great popular delusion" surrounding Martian inhabitants would never have occurred "if Schiaparelli had used any other modern language" but Italian.[1] This chapter shows, however, that it was the cartographic *image*, not the term, which spurred a furor over the Martian canals. The processes and inscriptions of scientific cartography allowed partial and uncertain observations of Mars to become established as objective astronomical truths. So strong was the correlation, in fact, that those truths began to evaporate the moment the astronomical map lost its status as a proper scientific representation.

Putting Mars on the Map

Before 1840 astronomers typically recorded their views of Mars in abstract sketches. With good views of Mars proving exceedingly rare, it was not known whether observed surface markings should be interpreted as surface features or atmospheric cloud patterns. Even as the sketches increased in complexity from the seventeenth to the early nineteenth century, this uncertainty remained. Sketches of Mars therefore typically retained the convention of representing the aspect of Mars at a specific and unique moment, as seen by a specific and unique viewer. This type of representation implicitly acknowledged that different viewers might possibly see very different images when looking through the telescope at Mars, even at the exact same time.

By 1840, however, sufficient certainty that astronomers were gazing upon the unchanging surface of Mars (rather than upon a swirling veil of clouds) allowed German astronomers Johann Mädler and Wilhelm Beer to produce the first modern "map" of Mars (see fig. 2.1). Their cartographic representation was constructed on a planar (azimuthal) projection that visually prioritized the Martian polar regions, which had become areas of major interest to astronomers, due to the apparently seasonal waxing and waning of irregular white patches at both the north and south poles. Beer

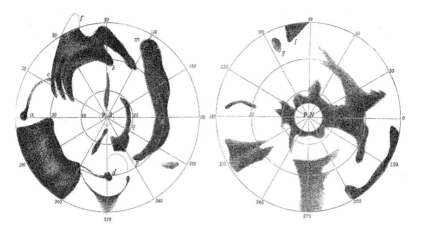

Figure 2.1. Planar projection maps of Mars by German astronomers Johann Mädler and Wilhelm Beer, 1840. These are the first known representations to use definitive labels for Martian surface features and to place them within a latitude/longitude coordinate system. This version appeared as figure 68 on p. 107 in Flammarion, *La planète Mars*.

and Mädler's representation departed from the sketch tradition in two important ways. Theirs was the first published image of Mars that was based on an extensive compilation of numerous observations, sketches, and measurements rather than single views from specific moments and places. In addition, it depicted a number of definitively marked and alphabetically labeled features on top of a grid coordinate system that was anchored by a prime meridian. When Martian features were suspected to be vaporous or changing, individual and subjective sketches had been considered appropriate representations. The use of geometric cartography, however, indicated a new certainty that the same features would appear in the same places again and again, allowing for an objective rendering of their locations and shapes. Although individual sketches of Mars continued to be published after 1840, this map signaled the start of a new era in the scientific representation of Mars. By the late nineteenth century cartography had become the primary mode of representing scientific data and knowledge about Mars.

Throughout the 1860s and 1870s, regular Mars observers began producing their own maps or at least forwarding their sketches to other astronomers who were known to be producing maps. Within the British astronomical community, for instance, leading planetary observers distributed standardized sketch sheets to their colleagues, provided detailed instructions on observation and drawing techniques, then collected contributors' notes and sketches for compilation into lengthy reports and detailed maps at the end of each biennial opposition.[2] (An "opposition" occurs when Mars and the sun appear on directly opposite sides of Earth.) These maps, born of collaborative effort and standardized practice, essentially removed any personal identities and subjectivities from the visualized geographic data. The new cartographic view of Mars thus assumed a powerful authoritative claim to objectivity.

In the realm of popular science, cartography made its debut in the late 1860s, when English astronomy writer Richard Proctor produced two comprehensive maps of Mars for book publication. A believer in both natural theology and pluralism (the existence of life on other planets beyond Earth), Proctor considered Mars to be a good example of his argument that God had created all of nature for the purpose of supporting higher forms of life.[3] In writing extensively about the red planet and its certainty of hosting intelligent life, Proctor relied heavily on maps to help make his points. Basing his maps on sketches made by the respected English astronomer William Dawes, Proctor created a comprehensive cartographic view of Mars that showed the planet in great detail (see fig. 2.2). Amid growing discussion of whether and to what extent Mars might be like Earth, Proctor used the perceived objectivity and authority of the map to make a very strong case for analogy between the two planets. For instance, he replaced Beer and Mädler's alphabetical labels with proper names formed by combining astronomers' surnames with generic designations such as "bay," "ocean," and "continent." Not only did this nomenclature refer to specific forms of terrestrial terrain, but the map also implied a Mars–Earth analogy in the very fact that it showed the Martian landscape to be mappable and nameable after the same manner as Earth's own landscapes.

Although Proctor's books were written for popular audiences, his maps carried distinct authority in the scientific world as well. Lightman has shown that the two projections Proctor used for his published maps carried different and complementary types of authority.[4] The stereographic projection Proctor used for a Mars map published in the widely read *Other Worlds than Ours* (see fig. 2.3) would have been familiar to Victorian audiences that considered stereographs to be reliable and truthful representations of nature. Furthermore, the circular representation of each hemisphere would have been easily compared to the source data—Dawes's circular sketches, some of which were published in the same volume—thus reinforcing the perceived scientific accuracy of the map. Proctor's later use of the Mercator projection (see fig. 2.2), however, created a wholly different view of Mars, despite showing essentially the same landforms. The Mercator projec-

Frontispiece.

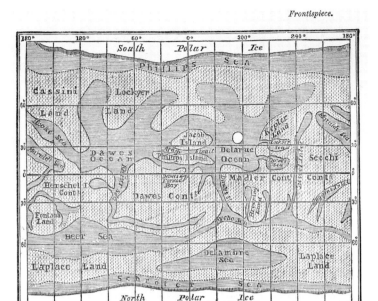

CHART OF MARS ON MERCATOR'S PROJECTION.

Figure 2.2. Mercator projection map of Mars by English astronomy writer Richard Proctor, 1869. First published in *Popular Science Review*, this map also appeared later as the frontispiece in Proctor's 1872 book *The Orbs Around Us*.

tion, which was first conceived and used as a navigation map, had become a standard world-map projection by the nineteenth century, even though it severely distorts upper latitudes and polar regions.[5] By using this projection, Proctor visually prioritized the Martian equatorial regions and deepened his implication that Mars was essentially mappable, navigable, and controllable in the same way that the imperial powers had come to see the equatorial regions of Earth.[6] After Proctor's Mercator, which was published in 1869 and again in 1872, planar projection maps of Mars (à la Beer and Mädler) became less common while the Mercator projection became the standard cartographic format for representing Mars data.

In addition, Proctor's nomenclature drew professional astronomers into several squabbles over the naming of Martian surface features. Once the Martian landscape had been labeled with proper names, it might seem a matter of simple convenience for all other astronomers interested in Mars to adopt those same names. Having disproportionately filled the map with English astronomers' names, however, Proctor provoked a display of nationalistic territoriality among amateur and professional astronomers throughout the

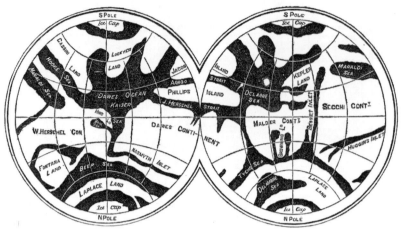

A Chart of Mars, laid down on the Stereographic Projection, by R. A. Proctor.
From Drawings by Dawes.

Figure 2.3. Stereographic projection map of Mars by English astronomy writer Richard
Proctor, 1868. First published in *Fraser's Magazine,* this map also appeared later in two of
Proctor's influential books: *Half-Hours with the Telescope* (1868) and *Other Worlds than Ours*
(1870). This version appeared on p. 109 in the 1900 edition of *Other Worlds than Ours.*

European continent. The well-known French astronomer and popular science
writer Camille Flammarion, for example, altered Proctor's scheme for his
own maps, adding more continental astronomers' names and thus supposedly remedying the undue favoring of Englishmen in Proctor's original
map.[7] Despite Proctor's insistence that he had applied astronomers' surnames rather casually and did not mind if another naming scheme were
proposed and adopted, his choice of cartographic labels exposed significant
clefts within the scientific community that soon deepened into major rifts,
as discussed later in this chapter.[8]

Perhaps the most significant influence of this popular cartographer on
scientific astronomy, however, can be seen in the undisputed acceptance of
his conventions of using astronomers' surnames and of delineating proper
names in the first place. As with his choice of the Mercator projection, Proctor's decision to apply a surname-based nomenclature to Mars implied that
it was an Earth-like world that could be known, mapped, and dominated in
the same ways as Earth's spheres of imperial expansion and colonial control.
The persistence of these basic elements of Proctor's cartographic approach
indicates that his underlying belief in a critical analogy between Mars and
Earth was widely accepted, even if his specific claims or representations met
with some opposition along the way.

The Maps of 1877–1878

The year 1877 marked a turning point in the cartography of Mars. On September 5 of that year, Earth and Mars stood in "perihelic opposition," as Earth came into a line between Mars and the sun at a moment when the two planets were each nearest the sun and also to each other along their respective elliptical orbits.[9] With the disk of Mars fully illuminated by the sun during this close approach, terrestrial astronomers enjoyed incomparable views, not only on the day of the perihelic opposition, but also in the days and weeks leading up to and following the actual event. Taking advantage of this rare occurrence, many astronomers observed Mars more extensively and with more interest than usual. In the year after the opposition, two very different maps were published by astronomers who observed Mars in different ways and from different places. As presumably objective records of scientific data, these Mercator projection maps made implicit and authoritative claims about what the surface of Mars was like. Given the stark differences between the maps, however, only one of these claims could be considered "correct." This section examines the scientific debates and competitive maneuvering that eventually determined which claim was superior, showing that the contrasting production methods of the two maps played almost no role. Instead, the visual authority of markedly different mapping styles, supplemented by the personal authority of the individual mapmakers, eventually determined which of the two maps would become more influential.

In the summer of 1877 English amateur astronomer Nathaniel Green departed from his usual observing station—in the back garden of his home in St. John's Wood, a suburb of London—and voyaged to the Portuguese island of Madeira in search of good atmospheric conditions for extended observations.[10] Over two months, Green's effort was rewarded with forty-seven nights suitable for Mars observation, sixteen of which he termed "good," "excellent," or "superb." This was less than Green had expected but still "considerably in excess of the average of an English climate." While sitting at the telescope on these forty-seven nights, Green recorded his impressions of the Martian landscape in forty-one color sketches, each of which took approximately two hours to prepare.[11] (See, for example, fig. 2.4.) Twelve of these sketches were published in his lengthy observation memoir, along with an orange-toned Mercator projection map and planar projection maps for the Martian north and south poles (figs. 2.5 and 2.6). Green's foreign expedition and exquisitely detailed map cemented his status as a serious and capable amateur astronomer.[12]

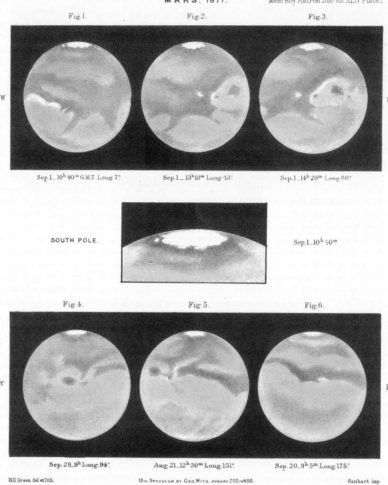

M A R S. 1877.   Mem.Roy.Astron.Soc.Vol.XLIV.Plate I.

Figure 2.4. Sketches of Mars recorded at the telescope by English astronomer Nathaniel Green, 1877. Green published sketches from his logbook to support the accuracy of his Mercator and planar projection maps, which included detail from his own as well as others' observations of Mars. These sketches originally appeared in color (using reddish-orange tones) as plate 1 in Green, "Observations of Mars." Courtesy Richard McKim.

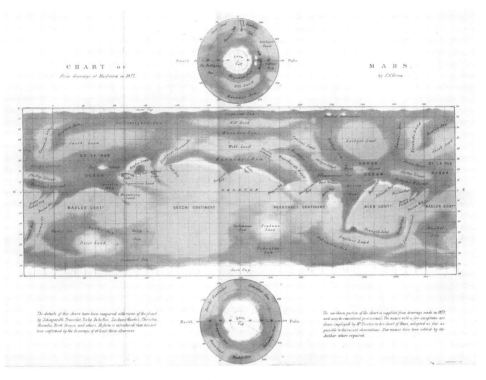

Figure 2.5. Mercator and planar projection maps of Mars by English astronomer Nathaniel Green, 1878. Green used an orange-toned color scheme and naturalistic shading style to record his 1877 observations of Mars. Note that the southern hemisphere is shown at the top of the map, following the standard convention of mapping celestial bodies as viewed through the telescope. This map was published in Green, "Observations of Mars," which did not appear in print until 1879. Courtesy Richard McKim.

The map itself, published in 1879, included data collected throughout the 1860s and 1870s by Green and other observers in his large network of British colleagues.[13] As such, the map was meant to represent the sum total of British astronomers' knowledge about Mars, rather than the individual views that Green recorded in Madeira. Although Mars' northern latitudes were not visible from Earth in 1877, for instance, Green's map covered all latitudes from 80° south to 80° north, based on previous data. Green made much of the comprehensive nature of this map, publicly seeking little acclaim for his own efforts. In vouching for the accuracy of his published map, in fact, Green claimed it included no major markings that had not been definitively seen in at least three separate observations, even leaving out prominent items that some of his colleagues insisted should have been

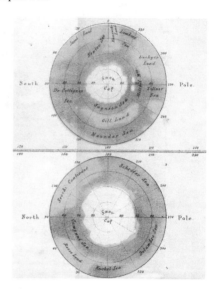

Figure 2.6. Planar projection maps of the Martian polar regions by English astronomer Nathaniel Green, 1878. Green's 1878 maps included data from the entire surface of Mars, even though its northern latitudes were not visible from Earth during the 1877 opposition. To provide this comprehensive view, Green relied on data that had been collected by other observers at previous oppositions. These maps, which were originally published in color using orange tones, have been enlarged from the plate shown in fig. 2.5. Courtesy Richard McKim.

included.[14] Green's personal contribution to the map, aside from its celebrated naturalistic rendering, was an augmentation of the detail visible in Mars' southern latitudes and the addition of several new place names. For instance, he added to Proctor's nomenclature with the honorary designation of "Schiaparelli Lake" for a feature he discovered during his Madeira observations.

The name for Green's newly discovered lake was intended to acknowledge the efforts of another astronomer who produced an influential map after the 1877 opposition. A professional Milanese astronomer, Giovanni Schiaparelli, had not intended to make more than casual observations of Mars in 1877, but he soon became intrigued by the planet's complex appearance during its close approach. Although he was essentially a first-time Mars viewer with no reputation for planetary work, Schiaparelli conducted a detailed study of unprecedented length that cultivated widespread interest within the astronomical world. Whereas most observers typically observed Mars for the few weeks just before and after opposition, Schiaparelli took detailed measurements of the planet's rotation and examined its markings for nearly eight months, including seven months after the opposition. Working from the Brera Observatory's rooftop telescope in Milan's stately Palazzo di Brera, Schiaparelli observed Mars from August 1877 to April 1878.[15] His logbooks include thirty-one complete drawings of Mars' face and more than 100 detailed sketches of various regions that he recorded during fleeting instants of "excellent air."[16] (See, for example, fig. 2.7.) Many of these pencil

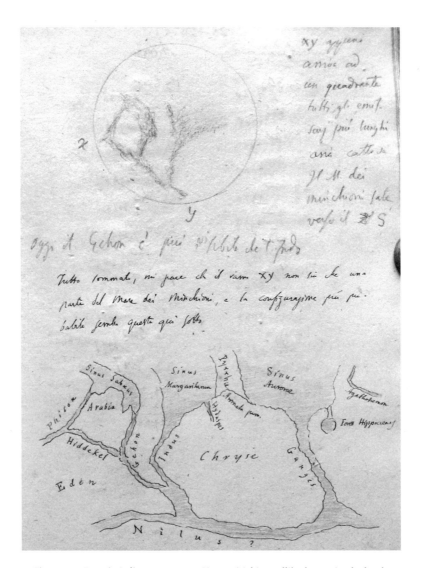

Figure 2.7. Entry in Italian astronomer Giovanni Schiaparelli's observation logbook, February 28, 1878. On the top half of the page, Schiaparelli completed a pencil sketch of the most prominent features appearing on the entire visible disk of Mars. The bottom half of the page includes a more detailed sketch (in ink) of a well-seen region, labeled with a classical Mediterranean nomenclature.
Source: L'Archivo Storico dell'Osservatorio Astronomico di Brera, Milan, Italy.

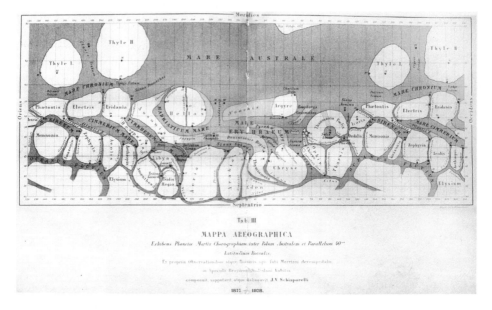

Figure 2.8. Mercator projection map of Mars by Italian astronomer Giovanni Schiaparelli, 1878. This map appeared as a plate in Schiaparelli's first Mars observation memoir ("Osservazioni Astronomiche"), which was published by the Royal Academy of Lincei in 1878. Schiaparelli depicted the northern hemisphere of Mars (shown in the lower half of the map) as a landscape of islands and peninsulas divided by narrow waterways. The dark tones on the map were originally printed in color, using a bluish-green tint. *Source*: L'Archivo Storico dell'Osservatorio Astronomico di Brera, Milan, Italy.

sketches were later tidied into composite drawings that Schiaparelli sent to colleagues, including Green, for comment. The full report of Schiaparelli's 1877–78 observations—including a colored Mercator projection map (fig. 2.8) as well as a planar projection map of Mars' south pole (fig. 2.9)— was published by the leading Italian scientific society in 1878.[17]

Schiaparelli's map was radically different from and seemed to contradict Green's effort. Where Green had used subtle naturalistic shading to represent an orange surface mottled with barely perceptible "delicate markings," Schiaparelli had used hard-edged black lines to show a detailed landscape of white islands divided by parallel and intersecting blue straits he labeled "canali."[18] Numerous stark features appeared on his self-made map that were nowhere to be found on Green's compilation, despite the fact that Schiaparelli's map included almost no detail north of 40° latitude due to its invisibility from Earth in 1877–78.[19] Schiaparelli's new map also rejected Proctor's surname-based nomenclature scheme outright, replacing it with a set of Latin place names based on the classical and mythological geography

Tab. IIII.

Hemisphærium Martis Australe

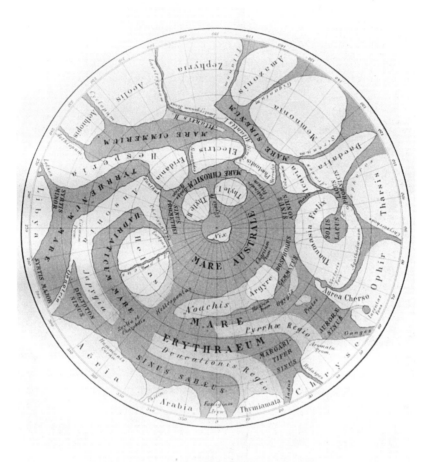

stereographice descriptum.

Figure 2.9. Planar projection map of Martian south pole by Italian astronomer Giovanni Schiaparelli, 1878. This polar view accompanied the well-known Mercator map (fig. 2.8) that appeared with Schiaparelli's observation memoir in the 1877–78 volume of *Atti della Reale Accademia dei Lincei*. As with fig. 2.8, the dark tones originally appeared in a bluish-green tint. *Source*: L'Archivo Storico dell'Osservatorio Astronomico di Brera, Milan, Italy.

of the Mediterranean world. "Lockyer Land," for instance, was renamed "Hellas," while "Fontana Land" became "Elysium."

Schiaparelli's map was a fundamental departure from those that had come before, and it was largely irreconcilable with the map Green had produced in the same year. Schiaparelli tried to assuage the skepticism he presumably expected by focusing on the extent to which his 1877–78 observations agreed with those that had been made by prior astronomers. Throughout his memoir, he referred repeatedly to the work of respected Mars observers from the past and claimed that the detailed observational data recorded in his map were merely "confirmations" of features that had already been seen:

> Many configurations, which judging superficially by my chart might appear as new, are found to have been described at earlier times, with greater or less evidence; while many details of the previous sketches, of which it has been difficult or impossible to be certain, are confirmed from my observations in this way. It is this mutual confirmation of results, more than the discovery of new details, which in my judgment provides the utility of our essay in areography.[20]

Although Schiaparelli could claim neither a long personal history of observing Mars nor a reliance on the data of Mars-observing colleagues, as could Green, this textual strategy was meant to temper the radical nature of his new map.

The contradictory appearances of the two new maps, however, were not so easily accepted. Both Green and Schiaparelli made similar claims as to the objectivity and accuracy of their Mars data. Both astronomers discussed the power and exactness of their telescopes, the unique atmospheric clarity at their observing locations, the first-hand "eyewitness" quality of their observations, and the essential agreement of their own sketches with the work of earlier observers. Additionally, both astronomers cast themselves as objective, unbiased observers, as in this claim of Green's: "Each drawing was made direct from the telescope, and entirely independent of those which had been produced previously; all comparisons being reserved till the evening was over, so that each view might be as free as possible from bias, or a leaning towards the repetition of similar forms."[21]

Given these claims to objectivity and accuracy, however, both maps could not be correct, which both astronomers acknowledged, however politely. Green expressed surprise at the draft map that Schiaparelli sent him, as Green claimed to have seen no prominent lines such as those that appeared on the Italian astronomer's map. Tactfully noting that the two maps otherwise concurred, however, Green suggested that the discrepancy could perhaps be chalked up to differences in draftsmanship. At a meeting of the Royal Astronomical Society in April of 1878, Green shared a series of pre-publication sketches that Schiaparelli had sent to him, saying he "hoped he should be excused if he exercised a little artistic criticism on the drawings. He thought the hard and sharp lines must be an error, and were the result of some process which Prof. Schiaparelli had adopted in making the drawings."[22] Similarly, in a personal letter to Schiaparelli, Green wrote that he was "much pleased to find that there is so much agreement in the large and general forms between [the drawings made at Milan], and the series I have made at Madeira. We evidently intend the same thing though we have a different way of expressing it." Schiaparelli did not respond in print, but expressed private displeasure at what he perceived as Green's "thoughtless" initiation of a controversy between them.[23]

Although Schiaparelli made little public effort to defend himself against such unflattering characterizations, his map quickly gained the upper hand in the scientific community. Green's argument about artistic style notwithstanding, many astronomers apparently concluded that Schiaparelli had truly seen something new on the surface of Mars. His map essentially touched off a canal-hunt, with scores of professional and amateur astronomers across Europe and North America committing themselves to the challenge set out by Schiaparelli's ally, the Belgian astronomer François Terby, to "verify the *positive* observations of M. Schiaparelli, whose chart if it were verified would constitute the greatest step made by areography for many years."[24] Green himself wrote before the 1879 opposition that "a careful search should be made for the remarkable dark canals figured by Professor Schiaparelli" and asked British observers to forward their sketches to him for analysis.[25] Although Schiaparelli alone reported seeing significant numbers of canals in the oppositions of 1879, 1882, and 1884, his observations were finally duplicated by both Terby and the French astronomer Joseph Perrotin in 1886.[26] As if suddenly freed from the constraints of deference to Schiaparelli's British opponents, a wide variety of European astronomers began to see and map the Martian canals after 1886. In the ensuing three decades, most Mars maps produced in Europe and North America used Schiaparelli's nomenclature and artistic style.

In addressing the question of why Schiaparelli's unprecedented map—which contradicted almost every existing trend in Mars cartography and then went unconfirmed for nearly a decade—had such a dramatic and lasting effect, we must first note the difference in professional standing between

Schiaparelli and Green. Although Schiaparelli had not been known previously as a planetary observer (his major career discovery was the theoretical prediction and observational confirmation of the link between meteor showers and comet orbits), his impeccable academic pedigree, long list of publications, and successful directorship of Milan's Brera Observatory had established him as one of the leading astronomers in Europe.[27] As such, he was generally treated with deference and respect, even by those who were skeptical of his unorthodox map. In professional society meetings and publications throughout the 1880s, the European astronomical community revealed a willingness to entertain all manner of explanation for Schiaparelli's canals. Green thought the dark streaks might be artistic misrepresentations; Maunder considered them most likely to be the boundaries of differently shaded regions; and another writer for the British journal *The Observatory* suggested Schiaparelli might have been using too high a magnifying power for his telescope.[28] Green himself was at pains to make clear, however, that his critique of Schiaparelli's mapping style was not meant as criticism of the astronomer's talent as an observer. Although he enjoyed a prominent reputation in Britain, Green was an amateur observer and clearly did not outrank Schiaparelli within the discipline. Referring to Schiaparelli deferentially as "the learned and exact professor," Green justified his limited criticisms of Schiaparelli's map only on the basis of his own status as a professional portrait artist and drawing-master, restricting his comments to the artistic style of the maps.[29] At a meeting of the British Astronomical Association,

> [Green] began by remarking that the point he wished to raise was purely one of drawing, and not one of seeing. It was one thing to see a difficult marking; it was quite a different matter to represent it accurately and artistically, nor was it any reflection upon an astronomer's ability to call in question his powers of drawing. They had no right to assume, as a matter of course, that such ability would accompany his other attainments.[30]

Lord James Lindsay, president of the Royal Astronomical Society from 1878 to 1879, commented similarly, "Professor Schiaparelli was not likely to be led away by imagination. There might be something peculiar in his telescope, or in his eyes, but he was not likely to publish observations or drawings without being fully persuaded that the appearances actually existed."[31]

This deference to Schiaparelli's professional reputation, however, does not fully explain why Green's map never managed to achieve more authority within the scientific community. Although he was not a professional astronomer, Green was a well-known and accomplished amateur at a time when professional astronomers were not typically active in planetary observation. As evidenced by his eventual election as president of the British Astronomical Association in 1896, Green's reputation as a Mars observer and his status within the scientific community were unimpeachable. It was not purely professional reputation or personal prestige, then, that accounted for the two maps' different fortunes. Nor was it the substantial procedural differences that had been used in making the maps. As already noted, Green included the observations of other astronomers in his map, while Schiaparelli projected only his own sketches. Green spent hours on each of his sketches, while Schiaparelli dashed off details as quickly as they appeared and then refined the map later. The maps themselves concealed these differences, however, asserting a scientific authority separate from the identities of the mapmakers or the procedures of the mapmaking.

As numerous critical historians of cartography have now shown, maps are typically viewed as objective representations of reality, despite their inherently ideological underpinnings. Cartographic representation has therefore been extremely powerful in its ability to present certain claims as inevitable or incontestable.[32] As applied to the case of Mars, this cartographic power ensured that the inscription of detailed landscape features on a latitude/longitude grid would be seen as scientific certainty regarding the existence and nature of actual features of the planet's surface. In showing a greater level of detail with a definitive style of marking, Schiaparelli's map thus appeared to be much more certain than Green's hazy and indistinct colorations. Despite Green's objections that Schiaparelli's artistry and coloration were flawed, his own map faced the impossible challenge of demonstrating *more* authority by presenting *less* detail. Where Schiaparelli could claim to have seen something that no one else had seen—the canals—Green was reduced to claiming that he was very sure he had seen nothing of the sort. This argument simply did not make sense when applied to the interpretation of cartographic data. Even though Green was essentially confirmed over and over (regarding the absence of canals) for nearly a decade, it took only one confirmation of a canal sighting to fully legitimize Schiaparelli's map.

Battling for Martian Territory

In the intervening nine years between Schiaparelli's first canal sighting in 1877 and the confirmation of his canal observations in 1886, a complex competition developed over Mars knowledge. While astronomers across Europe and North America tried diligently to find Schiaparelli's canals, the question remained about what to do with his new place names. Some astronomers were inclined to accept his more detailed map but hesitated to change the entire nomenclature scheme. Others supported a fully de-Anglicized nomenclature but couldn't accept a map covered with canals. As this section shows, however, Schiaparelli's place names were intricately linked to his canals through the cartographic format. As a result, Schiaparelli's map provoked a much more intense version of the battles that took place when Proctor had labeled Martian landforms in the first place. A decidedly nationalistic and territorial battle to control the production of Mars knowledge focused on the map as a locus of scientific territory.

In using Latin names drawn from Mediterranean geography, Schiaparelli reinforced his map's visual effect of casting Mars as a familiar, Earth-like world. Where Proctor had implied a Mars–Earth analogy simply by evoking imperial cartography and its domination of foreign landscapes with the application of European surnames, Schiaparelli pursued a different route. He directly asserted not only a general analogy between Martian and terrestrial topography but also a specific analogy between the Martian landscape and various regions of Earth:

> The immense region which has received the name Ausonia extends a quarter of the way around the planet's globe, and shows in form and disposition a great likeness to the terrestrial land of Ausonia [Italy]; from this likeness is derived its name and also those of Eridania, Hellas, and lastly Libya, which forms the other land bordering the Tyrrhenian Sea.[33]

By using his place names to imply that the Martian landscape was similar to that of the Mediterranean world, Schiaparelli went much farther than Proctor in asserting the Mars–Earth analogy. At the same time, he subtly unburdened the map of its English connections and asserted instead a new link to southern Europe and the Mediterranean basin.

The removal of English astronomers' names to make way for Schiaparelli's Italy provoked a rather heated reaction. Many British astronomers found the new names silly and resented Schiaparelli's unilateral rejection of the existing nomenclature. When the editors of the British journal *Astronomical Register* asked readers in 1878 to submit their comments on the nomenclature of Mars, one British astronomer lamented that Schiaparelli's contribution had served only "to create wholly needless confusion," while another dismissed the Latin names as "useless rubbish."[34] Schiaparelli himself did not help the matter by making conflicting claims about his decision to relabel the Martian landscape. On the one hand, he tried to stay above the fray by claiming that the new nomenclature was based on personal whimsy:

> *I seek neither the collective approval of astronomers nor the honor of seeing it pass into general use.* To the contrary, I am ready to adopt later whatever scheme will be recognized as definitive by the proper authority. Until then grant me the chimera of these euphonic names, whose sounds awaken in the mind so many beautiful memories.[35]

At the same time, however, he claimed that the Mediterranean-geography names were a necessary product of analogy: "In general the configurations presented such a striking analogy to those of the terrestrial map that it is doubtful whether any other class of names would have been preferable."[36] As for the truthfulness of these two contrasting claims, Schiaparelli's logbook shows that the first features he sketched were immediately named for actual terrestrial locations, while the more symbolic and mythical names were filled in later. On September 11, 1877, Schiaparelli first recorded proper names in his observation logbook, referring to a "Mare Tireno" (Tyrrhenian Sea), "Adriatico" (Adriatic Sea), "Grecia" (Greece), and "Ellesponto" (Hellespont, or Dardanelles).[37] This chronology confirms that Schiaparelli's nomenclature reflected, first and foremost, a sense of real analogy with Earth's landforms. His published claims about the nomenclature's whimsical nature were therefore more likely efforts at scientific diplomacy.

Despite their objections, Schiaparelli's detractors could see no reasonable way to reclaim the map. The only immediate alternatives—derivatives of Proctor's surname labels—were admitted to be problematic in their prioritization of various individuals and nationalities over others. One writer commented, "It may be a *present* compliment, but must be simply ridiculous to *future* astronomers, to call each newly-discovered marking by the names of individuals of no lasting scientific eminence."[38] Nonetheless, many commentators seemed to agree that the names applied to Martian features should be decided by the mapmaker. When asked to make a statement on the controversy, for example, the esteemed Scots astronomer Sir David Gill responded that

> The question can only settle itself when, party feeling on the subject having been forgotten, a map of Mars, so superior to all others in convenience and accuracy, appears, that by its simple merits alone . . . it becomes a standard of reference without controversy. The matter, therefore, I think, should be left

to the judgment of the man who may be successful in producing a map that shall *command* the position of authority.[39]

In essence, then, the map itself became the repository of Mars knowledge and expertise, subject to both territorial and nationalistic maneuvering. As one British astronomer remarked during the debate over Martian place names, "the discovery of any fresh areographical feature renders it, in one sense, a portion of the scientific possessions of the nation in which it may happen to be made." As a record of such possessions, proposed maps of Mars had become infinitely more contestable than when Beer and Mädler had first inscribed Martian features on a coordinate system of latitude and longitude. At one level, it was hard for some astronomers to envision engaging in aggressive territoriality over such a distant world: "We are in the last degree unlikely to go to war either with the Belgians or the Italians to obtain a 'scientific frontier' in Mars and I myself cannot see any valid objection to Cape Schiaparelli, or to Terby Sound, upon a map of the planet."[40]

At another level, however, the British *did* go to war with continental Europe over Mars. In struggling to control the map's nomenclature and protect British prestige, many British astronomers conducted a war of words that functioned in many ways like a classic contest for territorial control. By Schiaparelli's own admission, control of place names was equal to control of the landscape: "The existing nomenclature simply proved insufficient for the vast quantity of new objects that had somehow to be named. . . . My nomenclature, which was devised at the telescope . . . is preserved in this memoir only because it describes perfectly what is seen"[41] With the canals and place names thus jointly inscribed on the map, any attempt to dispute one necessarily required removal of the other. If the names were merely a matter of preference or aesthetics, they could be changed or forgotten easily. If they were objective descriptions of "what is seen," however, they could not be replaced until a more accurate system was offered.

British astronomers' respectfully worded sniping about Schiaparelli's artistic ability and heated objections to his de-Anglicized nomenclature thus sought to protect Green's status as an equal discoverer of Mars' southern features. If not for the explanation of the maps' differences on the basis of artistic style, Green might have been forced to admit that Schiaparelli saw more, saw better, or saw first, thus devaluing his expedition to Madeira. The failure of heated objections to Schiaparelli's nomenclature, however, only solidified the authority of the canal-covered landscape and allowed Schiaparelli to retain "discoverer" status for the apparently new features.

Long after the 1878 nomenclature debate had ended, many British astronomers stubbornly held on to the place names Green had used, resorting to Schiaparelli's version only when there was no alternative. Twenty years after Schiaparelli's nomenclature was first proposed, for instance, a Mars report in the *Monthly Notices of the Royal Astronomical Society*, reported that "the Kaiser Sea has recently actually encroached upon the continent nearly so far as Lake Moeris, so as to obliterate part of Libya."[42] Such creative amalgamation was due in large part to nationalistic pride, as this retrospective comment in the 1900 *Journal of the British Astronomical Association (JBAA)* reveals:

> The only reason I can see for this attempt to discard the old names is that they were of English application, and so hurt the self-love of all who are not English. At any rate the selection of new names seems to have been made on the principle that no English need apply, and to be influenced by the same antipathy that makes our friends across the Channel desirous of removing the initial meridian to pass through Jerusalem or the Canaries, or in mid-ocean (because water is a more stable element than land), or anywhere so it does *not* pass through Greenwich.
>
> The names chosen are in many instances of unnecessary length, causing us to have to write or pronounce four or five syllables where two or three would suffice. And they are a remarkably evil sounding lot. They always remind me of the old lady who found Nebuchadnezzar or Beelzebub such a comforting word.[43]

Aside from reflecting a lingering British bitterness toward Schiaparelli's nomenclature decades after the new names had passed into general use, such statements also reveal that the competition over Mars' place names was every bit as nationalistic as other scientific competitions of the day. The explicit comparison of Mars debates with the contentious British-French argument over the location of Earth's prime meridian shows that nationalistic territorialism over scientific standards was a major motivation for British opposition to changes in the Mars map.

In addition to provoking such deep-seated territorialism, the newly inscribed names and canals conveyed a sense of analogy and intrigue for Mars that had not existed previously. Although Mars' dark features had long been referred to as "seas" and its light patches as "lands," the map's assertion that Mars boasted a "Libya," an "Arabia," a "Zephyria," and a canal named "Atlantis" cast Mars as a familiar, Earth-like world. And the fact that the map of this world had undergone a long (if nonviolent) siege only reinforced more strongly the conceptual acceptance of Mars as a geographical and territorial entity—a real world that could be delineated and contested by Europeans.

Projecting Authority

By the early 1890s, the authority of Schiaparelli's map had succeeded in legitimizing the canal-covered Martian landscape. First bolstered by the personal and professional reputations of various astronomer-mapmakers, increasingly detailed Mars maps themselves soon began to return the favor, conferring influence and status on their makers. In the context of self-replicating cartographic authority, maps of Mars became increasingly abstract in their focus on recording more and more canals. Even in the absence of a single "standard" Mars map, a remarkable trend toward increasingly geometric canal representations dominated the development of Mars cartography through the turn of the century. These abstract maps eventually became powerful cartographic icons that were viewed as indications of intelligent Martian life. This section examines the processes of cartographic projection, production, and authorization that supported and enabled this trend.

Throughout the 1880s, Schiaparelli continued to make updated maps that became increasingly abstract and geometrical. After his canal observations were confirmed in 1886, many astronomers began to follow his lead, reporting numerous canals at the end of each biennial opposition. As these newly sighted canals were added to increasingly complex maps of Mars, several astronomers struggled to explain the perplexing markings as natural features. Burton thought they might be "furrows" plowed by meteors striking the planet at oblique angles but was stumped by the number of lines that seemingly converged at circular intersections. Proctor suggested they were rivers, but was at a loss to explain why rivers would be so straight or how natural topography could explain so many intersections. Pickering claimed that the lines were more likely geological fissures formed by the rapid cooling of Mars' surface that released sufficient heat and water vapor to support strips of vegetation. None of these physical explanations gained any significant foothold in the literature, however, because they lacked a truly convincing terrestrial analogy.[44]

If the canals had not been formed by natural causes, however, the only other explanation must involve some "artificial" mechanism. In general, scientists found themselves fairly divided on this issue. Most astronomers accepted that the canals were real, given the frequency of their sighting at observatories throughout Europe and North America after 1886, but balked at any artificial explanation.[45] Some, however, began to consider that the canals might be illusory and therefore in need of neither natural nor artificial explanation.[46] Still others fell somewhere between these two positions, worrying that the seductions of artificial explanations had somehow come to compromise the observations themselves. For his part, the original canal observer Schiaparelli did not assert that the lines on Mars were of synthetic origin, but neither did he reject the idea. In a comment that was frequently quoted by inhabited-Mars advocates, Schiaparelli wrote in 1893: "It is not necessary to suppose [the canals] the work of intelligent beings . . . [but] I am very careful not to combat this supposition, which includes nothing impossible."[47] With this dissension among the professional cadre, popular audiences found little bar to their embrace of the artificiality hypothesis.

The optical-illusion theory, in fact, was at a strong disadvantage because Mars data was recorded in maps. Given the authority and nature of the cartographic data-recording format, it was nearly impossible to erase canals that had been mapped by a credible astronomer. Just as was true for many of the terrestrial expeditions of the day, prestige was derived by putting things *on* the map, not taking them off. Those who claimed to see a canal-free landscape on Mars did not even bother to produce or publish maps, as the reduction of detail was not considered a contribution of any importance. Astronomical maps thus functioned much like the self-replicating terrestrial maps of the day. British explorers such as Henry Morton Stanley, who added numerous features to the map of Africa, were hailed as heroes and began to set the agenda for British interests on that continent.[48] Those whose expeditions failed to turn up anything new, on the other hand, were branded failures and had difficulty finding sponsors for subsequent travels. Similar to the terrestrial explorers, Mars astronomers felt the need to include details from earlier maps in order to assert their legitimacy, even when those features could not be independently confirmed.[49] The use of pre-existing maps essentially as "background" images or platforms for the depiction of new canal data thus operated in many ways similarly to the now-notorious commercial maps of West Africa that repeatedly showcased the "Kong Mountains" over more than a century, even though the unconfirmed and nonexistent mountain chain continually reappeared solely due to the practice of copying detail from earlier maps.[50]

By century's end, an explosion of post-Schiaparelli canal sightings had given rise to maps resembling spiders' webs in their complexity. By the 1890s, geometric maps had become the standard representation of Mars, while any detailed rendering of shadings and colors was lost in the competitive quest to find and map new canals. Though Schiaparelli had apparently

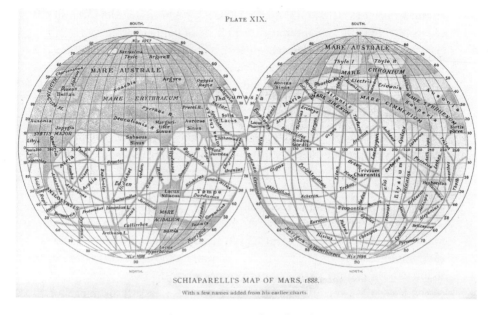

SCHIAPARELLI'S MAP OF MARS, 1888.
With a few names added from his earlier charts.

Figure 2.10. Stereographic projection map of Mars by Italian astronomer Giovanni Schiaparelli, 1888. As Schiaparelli observed Mars throughout the 1880s, he added more and more detail to his maps, which became increasingly geometric over the years. This version of his 1888 map was published as the frontispiece for *Astronomy and Astro-physics* in October 1894.

taken some of Green's original stylistic criticisms to heart, once sending his publisher a copy of Green's 1877 sketches with instructions to match the style and color tones for his own sketches, he persisted with the inclusion of definitive canal markings on his composite maps.[51] By 1888 the islands and channels of Schiaparelli's original chart had all but disappeared, replaced by thin lines that appeared inexplicably doubled in places (see fig. 2.10). The same trend took place in other astronomers' charts as well. New maps published every few years visually prioritized the representation of new canals to such an extent that, by the 1890s, maps of Mars consisted mainly of black lines and circles on a white background, with the names of various canals taking more prominence on the map than any subtle shading.

These abstract maps became strongly linked with the inhabited-Mars theory largely due to the efforts of American amateur astronomer Percival Lowell, who argued in his first articles and books about Mars in 1895 that the planet's "unnatural" and "artificial" appearance indicated the probability of intelligent life. Lowell's success in gaining support for this argument owed much to his active publication strategies but also relied heavily on the visual authority of his maps. Within an established competitive and territorial framework, those astronomers who added the most detail to the map became its most authoritative interpreters. Although Lowell had begun his Mars research with no professional pedigree, he quickly became one of the most prominent theorists about the landscape and culture of Mars by producing extremely detailed maps. Having funded the establishment of an observatory designed solely to observe and map the surface of Mars, Lowell immediately made an impact through his cartography. In his first year of observation, 1894, he confirmed all but two of Schiaparelli's canals and added 116 of his own discovery to the most detailed Mars map yet produced (see fig. 2.11).[52] Through his career as a Mars observer, Lowell continued to add considerable detail to his canal maps.

Whatever other astronomers might say about Lowell's speculative hypothesis, they had to admit that he deserved respect on the basis of his continued contributions to the Martian map. Simon Newcomb, director of the American Nautical Almanac Office and a noted Lowell antagonist, wrote to Lowell in 1905 to request a map for an encyclopedia article he was then preparing: "I would like a good map of Mars to accompany the article. For

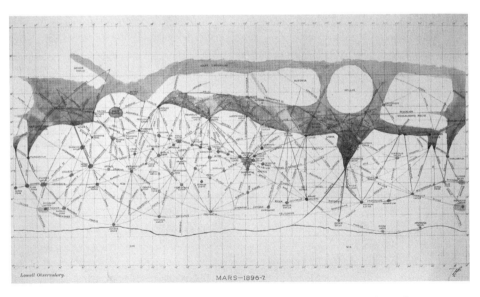

Figure 2.11. Mercator projection map of Mars by American astronomer Percival Lowell, 1897. Toward the end of the nineteenth century, geometric maps of Mars tended to visually prioritize the planet's presumed canal network. This abstract map of Mars did not appear in published form until 1905, owing to publication delays with the *Lowell Observatory Annals*, but Lowell had circulated it widely among other astronomers, many of whom referred to it in their own publications. Courtesy Lowell Observatory Archives.

this I know no better source than the publication of your observatory."[53] The editor of *Popular Astronomy*, W. W. Payne, likewise commented in 1904 that Lowell's maps were "pieces of astronomical work that are now classical in astronomy . . . because they were made by the very best means and methods now known to that science."[54] This comment probably owed more to the detailed appearance of Lowell's maps than to the actual process he used for mapmaking. In the popular press, praise for Lowell's attainments was even more glowing, as in an article that credited "the most interesting theory of all, the presence of life on Mars" to Lowell, "than whom no astronomer has made more important explorations to the other places in the Cosmos."[55] Whereas Schiaparelli's professional reputation had helped establish the authority of his canal-map, Lowell's legitimacy was produced by an opposite process: the unrivaled detail of his authoritative canal-maps actually produced significant personal authority that was not available to other amateurs.

By the early 1900s, Lowellian images of Mars had become powerful icons. In books, pamphlets, magazines, and newspapers, the gossamer network of the map—with its interlinked geometry of perfectly straight lines meeting at perfectly round intersections—became a ubiquitous symbol of extraterrestrial life. Popular Sunday papers, which published frequently about Mars in opposition years, almost always used geometric images of Mars to accompany their articles about the most recent astronomical discoveries. Though these images assumed the general appearance of the scientific canal-maps, they were often unlabeled or did not show any coordinates (see fig. 2.12). Such generic abstraction indicates that this iconic cartographic imagery was meant to convey legitimacy rather than information.

The strength of the Mars icon as a visual symbol rested not only on the map's powers of inscription, authorization, and legitimization. It was also supported at a fundamental level by the creative power of the cartographic process, which had brought into existence a landscape quite different from that which astronomers reported seeing through their telescopes. Despite the widespread use of geometrical canal imagery, in fact, no astronomer ever actually saw or claimed to see an interlinked canal network while sitting at the telescope. The cartographic authority of the increasingly prominent Mars icon concealed the fact that the canal "network" was actually invisible to the eye. From Earth, the surface of Mars was (and is) notoriously difficult to see. Even under conditions of excellent "seeing" (a measure of the stillness and clarity of Earth's atmosphere), distant Mars shimmered tantalizingly, allowing only fleeting glimpses of its surface. Astronomers

Figure 2.12. Cover story featuring Mars, *St. Louis Post-Dispatch Sunday Magazine*, 1907.
In colorful artwork alongside a feature article about Mars astronomy, the Martian
surface was depicted with the abstract geometric style common to the
scientific cartography of the time. Courtesy Lowell Observatory Archives.

constantly complained about their inability to "hold" an image of Mars in
the telescope, as detail could be seen only in glimpses and flashes:

It must not be imagined that any drawing represents what the observer sees
the moment he looks through the telescope. Instants of exceptional seeing
flash out, here and there, at different spots on the planet. It is not till the same

phenomena repeat themselves in the same way, in the same place, a great number of times, that the observer learns to trust these impressions. One has to keep one's mind constantly at the highest pitch to catch and retain what the eye sees.

It is like looking at a Swiss landscape from a high Alp, with the summer clouds sweeping about one. Now the mist rolls away, revealing a bit of the valley, and shuts in again in a moment; while in some other spot the clouds break away, and disclose a jagged summit, or a portion of a shining glacier.[56]

To give a quantitative sense for the duration of these moments, the director of the British Astronomical Association's Mars Observing Section wrote in his observation report for 1909 that "a *glimpse* of an object does not last more than 0.3 second; a *short view* of an object lasts from 0.3 to 1 second; and an object *held steadily* is one whose visibility continues for 1 second and above."[57] In essence, then, the art of sketching Mars consisted of waiting intently for a moment of still air, then quickly recording an image before the memory could fade. Given this difficulty, several astronomers insisted that a given feature should be seen, sketched, and measured multiple times before it could be definitively said to exist. Otherwise, the opportunity for mistakes—of vision, memory, or depiction—was too great.

As a result, very few of the sketches that astronomers drew in their observation logbooks or on standardized sketchpads depicted more than a few Martian surface details at any given time. It was only in the process of gathering, compiling, and projecting dozens (or even hundreds) of individual sketches onto comprehensive maps that astronomers gave rise to the view of a geometrical Martian landscape. Schiaparelli's famous chart included details from dozens of sketches recorded in his 1877–78 logbooks. Green's charts and others published by the Royal Astronomical Society and British Astronomical Association typically compiled the work of at least a dozen observers in London, Edinburgh, and many far-flung corners of the British empire.[58] Lowell's influential maps of the 1890s and early 1900s, likewise, were made by plotting the details from hundreds of his own and his colleagues' sketches directly onto a wooden globe (fig. 2.13), which was then tilted to the proper angle and photographed before tracing the negative into a Mercator projection.[59] Thus, very simple sketches blossomed cartographically into complex and interlinked networks that had never been seen by any single individual or on any single night. In truth, then, the networked appearance of the canals owed its existence more to the cartographic process than to any reality on the Martian surface.

LOWELL'S GLOBE OF MARS, 1903. *Frontispiece*

Figure 2.13. Wooden globe of Mars produced at the Lowell Observatory, 1903. The Lowell Observatory recorded the results of each Mars observing season by plotting astronomers' observations on wooden reference globes, which could then be used for mapmaking. All of these globes are still held in the Lowell Observatory Archives, including this one from 1903, which was photographed for publication as the frontispiece in Edward Morse's 1907 book *Mars and Its Mystery*.

Though astronomers admitted that the maps showed a landscape invisible to the eye, the authority of the complex scientific map conveyed an objectivity that outweighed the simplistic sketches. Detractors' criticism of the inhabited-Mars theory on the basis of the maps' incongruity with the drawings seem only to have cast suspicion on the simpler drawings, rather than decreasing the legitimacy of the detailed maps.[60] Even the theory's great champion, Lowell, acknowledged that the process of projection created an unviewable view: "not a single piece of the chart resembles the actual presentation of any part of the planet at any time." Though this comment was intended primarily to rebuff criticism from those who were unable to confirm his maps' canals through their own telescopes, Lowell seems also to have acknowledged at times the more creative role of cartography in bringing his populated "oases" to life: "*When they are plotted upon a globe,* they and their connecting canals make a most curious network over all the orange-ochre equatorial parts of the planet, a mass of lines and knots." Lowell's one-time associate Pickering made a similar caution: "The maps of Mars look very artificial; but we must remember that they are composites of many drawings, such as are given in this article. All the canals shown on the maps are not seen at once; on the contrary, only a very few of them are visible on the same night." Use of a coordinate grid, however, indicated exactness and scientific objectivity; projection of multiple observations into a composite view conveyed unassailable comprehensiveness. As an artifact of projection, therefore, the geometrical image of Mars could not have existed or grown so meaningful except through the format and process of cartography.[61]

Photography and the Decline of the Martian Map

Tied as it was to the map, the inhabited-Mars theory enjoyed widespread support only as long as cartography itself was accepted as the most trustworthy and "scientific" representation of the red planet. The development of new photographic techniques, pursued aggressively by Lowell as a way of supporting and legitimizing his maps, ironically led to the declining importance of astronomical cartography after 1910. Although astronomers continued to publish their Mars data in maps, photography emerged as a more objective and legitimate way of recording the appearance of the Martian surface. This section chronicles the interactions between these two competing data representations, showing how cartographic authority waned in relation to the increasing use and importance of planetary photographs.

After a brief hiatus from his Mars studies between 1898 and 1901 due to illness, Lowell returned to publishing with a renewed vigor. He published several new maps early in the twentieth century, wrote three new books by 1909, conducted extensive lecture tours on the American East Coast and in Europe, and disseminated his findings to the popular press at every opportunity.[62] His success in reaching the mainstream dailies can be read in assorted grumblings that surfaced in astronomical journals. The *JBAA* lamented in 1906, "We had extraordinary reports in sensation-mongering newspapers on this side of the Atlantic to the effect that some American observer, in the course of his nocturnal vigils, had detected the Martians in the act of signalling to the inhabitants of the Earth." Five years earlier, the same journal had claimed, "The idea of opening communication with other planets, and with Mars as a beginning . . . has been fostered by the sensational rubbish of magazine writers, and the extravagancies of newspaper paragraphists." *Popular Astronomy* cautioned in 1907, "The literature about Mars in the current magazines is, some of it fanciful, some funny, some very mysterious," having already reacted strongly to Lowell-inspired reports in 1895: "It is a burning shame that such nonsense finds place in our best and greatest daily papers."[63]

At the same time Lowell became more outspoken in his claims about the landscape and civilization of Mars, he also became more critical of those who did not accept his theories, prompting many of the most prominent American astronomers and several professionals and amateurs in Britain to work actively to discredit him. To combat what they saw as Lowell's willful disregard for scientific professionalism and standards of proof, his detractors retaliated with a sustained effort to disrupt his popularity and undermine his legitimacy. In Britain, for example, the well-known Royal Observatory, Greenwich astronomer Walter Maunder began to write extensively about the likelihood that Lowell's maps were based on nothing more than optical illusion, provoking significant doubt among those astronomers who had never seen the canals clearly in the first place. At a June 1903 meeting of the British Astronomical Association, the comment was made that Maunder "had really cut away the ground from under the feet of those who thought they had been able to prove that there were canals. The onus of proof now lay upon those who thought the canals were there."[64] In the United States, elite academic astronomers including Simon Newcomb, Wallace Campbell, and Edward Barnard acted in concert to isolate Lowell from the scientific community, cast doubt on his claims, and minimize his publishing opportunities.[65] Like Maunder, several of these American astronomers questioned whether Lowell's maps and sketches were distorted by optical illusion.[66]

To counter the many charges being leveled against him, Lowell turned to photography for redemption. After Maunder's first attacks in 1903, Lowell helped pioneer a new method of planetary photography that could capture a clear image with only a short time exposure. When his assistant, Carl O. Lampland, succeeded in photographing Mars in 1905, Lowell quickly began publishing and circulating the images to rescue his reputation. For a time, this strategy worked. Despite being small and grainy, the photographs indeed contained some dark markings in areas where Lowell's maps depicted canals, indicating a confirmation.[67] In popular American magazines and newspapers, the photographs were embraced as conclusive. *Scientific American*, noting that "the photographic plate cannot lie," reassured its readers that "the canals, whatever be their nature and origin, cannot be mere subjective illusions on the part of certain observers, but have an actual existence as material formations on the surface of Mars." *McClure's*, likewise, asserted of the canals that the photographs had "forever dispos[ed] of the assumption of their illusory character." This general response even extended into normally skeptical quarters. At a June 1906 meeting of the British Astronomical Association, for instance, the president, A.C.D. Crommelin, stated that Lowell's photographs proved the "objective reality of the canals," temporarily reviving belief within the British astronomical community.[68]

In 1907, however, new experiments were carried out in the United States to test the possibility that optical illusion was at work in the Mars observations. An influential experiment conducted by Newcomb found that trained astronomers who were asked to draw what they observed when a small paper disc covered with irregular markings was held at a great distance almost invariably drew straight canal-like lines that did not actually exist.[69] This finding appeared to confirm Maunder's earlier work on optical illusion, thereby producing an immediate reverse sway in scientific opinion over the reality of the canals, despite Lowell's vigorous rebuttals.[70] In the face of what he perceived as an onslaught, Lowell mounted a high-profile photographic expedition to South America for the 1907 opposition, essentially staking his reputation on the new imaging techniques Lampland had developed since 1905. As British and American magazines and newspapers hyped the expedition, scientific and popular anticipation mounted.[71] (This expedition is treated in more detail in chapter 4.) When Lowell's photographer finally returned from the Andes with the negatives, however, the images proved a general disappointment.

Lowell claimed that the 1907 photographs dispelled all doubt regarding the existence of the Martian canals. Paradoxically, however, they actually contributed to his further loss of credibility. Typically measuring about half an inch in diameter on the negatives, each photograph showed far less detail than any of Lowell's elaborate maps. Although the photos could be said to confirm Lowell's simple sketches, showing some isolated lines on the face of Mars' disk, they could not be said to show a definitive canal network. On top of that, they were incredibly difficult to reproduce: they were drastically small at original size but became excessively grainy when enlarged (see fig. 2.14). Lowell agonized over the proper presentation of his photographs in the *Century Magazine*, even asking that they be "retouched" to show the canals better. Having paid a substantial sum for the images' copyright, however, the editor was in no mood to delay publication of the long-promised Martian canal photographs: "There is no time to retouch the photographic plates and we should consider it a calamity to do so, as it would entirely spoil the autographic value of the photographs themselves. There would always be somebody to say that the results were from the brains of the retoucher."[72]

To counteract his expectation that the "unedited" photographs would reproduce poorly, Lowell began sending negatives and prints to select astronomers in Britain in the calculated hope that these men would vouch for the photographed canals in their own publications and presentations.[73] This strategy produced some desirable results. BAA President Crommelin reported that his personal examination of Lowell's images showed twenty-two canals. Likewise, the director of the BAA's Mars Section commented in his report on the 1907 opposition that "[r]egarding the objectivity of the canals of Mars, there seems no necessity or room for doubt after the truly splendid photographic results obtained by Messrs. Lowell and Lampland."[74]

Despite this personal vouching, however, the fact remained that Lowell's photographs were not convincing in any of the formats available for mass distribution. They appeared too small, too blurry, or too dark to match the certainty levels that had been inscribed in the maps. Wherever the much-hyped photographs were published, Lowell usually insisted that a disclaimer accompany them. In the 1907 *Century* exclusive, for example, Lowell alerted readers that the printed images were three steps removed from the original negative, due to photographic printing, half-toning, and press printing. He further cautioned that use of a magnifying glass would only increase the grain size without revealing more detail. Lowell was thus forced to make a delicate argument. Asserting on the one hand that "to the camera no evasion of the fact avails. They [the canals] are there, and the film refuses to report them other than they are," he was forced on the other to qualify the photographs as "handicapped," claiming the canals' "straightness is more pronounced than appears from the photographic print."[75]

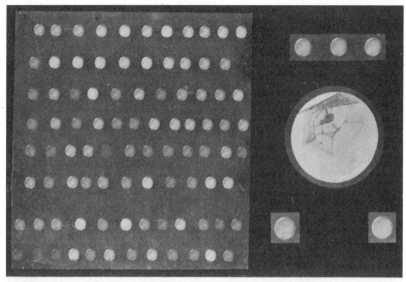

No. 4. SHOWING VARIETY OF INTEN-
SITY OF IMAGE. THE DIFFERENCE IS
DUE TO VARIATION IN TIME OF EX-
POSURE

REGION OF THE SOLIS LACUS. LONGI-
TUDE OF THE CENTER OF THE PHOTO-
GRAPH, 90°. ENLARGEMENTS, AND PRO-
FESSOR LOWELL'S DRAWING

Figure 2.14. Photographs of Mars published to illustrate Lowell's 1907 article in *Century Magazine*. This photographic display appeared on p. 308 in Lowell, "New Photographs of Mars." The small circles in the array at left were produced directly from a sheet of negatives at original size (with each hardly larger than a few letters of the magazine's text). The intermediate-sized circles on the right are enlargements of selected photographic originals centered on the 90-degree meridian. The largest circle (center right) is a sketch of the same region from Lowell's 1907 observation logbook, which he included for comparison and as a guide to the enlarged photographs. Lowell's text warned the reader, however, that the process of enlargement was of minimal use, as it also enlarged the grain of the photograph and "must not be overdone."

Perhaps more damaging than the inadequate reproduction of the tiny photographs, however, was the fact that photography supplanted cartography after 1907 as the proper standard of proof for Mars representations. The buildup of expectations regarding the Lowell photographs focused on their purely objective quality and their ability to resolve long-standing disputes among astronomers over the existence of the canals. Once the grainy photographs had been obtained, Lowell's elaborate maps—the basis of his reputation, credibility, and hypothesis—became nearly obsolete as scientific images. In a 1907 letter discussing the illustration of an article on Mars for the tenth edition of the *Encyclopaedia Britannica*, for example, editor Hugh Chisholm wrote to the author Simon Newcomb that he did not want to publish Lowell's maps or drawings:

I think that only a half-tone reproduction of Lowell's *photographs* would be scientific. . . . The whole thing in fact is so much bound up with the Lowell photographs that I shrink from *showing* anything but the originals (which are decidedly difficult for us to reproduce, and had better be therefore referred to only in their source). . . . I don't in any case like the idea of mere drawings, which must inevitably "fake" to some extent the "canals."[76]

In the end, Chisholm decided he would publish the encyclopedia's "Mars" entry with no image whatsoever, rather than use any cartographic stand-ins ("mere drawings") for the "scientific" photographs.

Many editors seemingly came to a similar conclusion after the vaunted 1907 expedition, as Lowell's maps rarely appeared in scientific publications after that year. Photography had provided a new imagery of truth that made astronomers' diverse maps appear positively subjective in comparison. The fact that the photographs were blurry and grainy did not diminish their perceived objectivity. It did, however, diminish the certainty of the canals that had been inscribed in Lowell's and others' maps.

With doubts cast on the authority of cartography as an objective format, astronomers were forced to confront the fact that photography was not yet capable of providing definitive answers about the geography of Mars. Grainy, blurry photographs, however authoritative and objective they were considered, could neither prove nor disprove the existence of canals or the correctness of the canal theory. Compared to stellar topics, where improved technology and techniques in spectroscopy and photography could provide definitive answers to exciting questions, planetary topics faded into the scientific background for several decades. At the same time, popular enthusiasm for Mars began to show its first signs of waning as well. Without definitive answers to questions such as whether Martians were signaling to Earth, public interest turned to other topics over time. Though it took much longer for popular interest in the Martian canals to die out (it arguably continued with some audiences into the 1950s, if not to the present day), the decreasing power of the map had a marked effect on both scientific and popular audiences' confidence in the supposed Martian inhabitants. Having risen to prominence as the most eloquent and active promoter of the inhabited-Mars hypothesis, it was Lowell who suffered most keenly from this decline of the map.

A Scientific End for the Canals

Another damaging blow to Lowell's scientific credibility came in 1909–10 when he became embroiled in a debate that bore striking resemblance to the old Schiaparelli-Green disagreement over whether Mars was best represented with hard-edged lines or naturalistic shading. With the authority of his map weakened by the new photographs, Lowell's credibility as a theorist or interpreter of Martian geography was also newly vulnerable. Whereas he had earlier been able to maintain a spirited defense against all criticisms, he was left after 1907 to argue from a much weaker position. Those astronomers who had long wanted to dismiss Lowell's theories and speculations regarding Martian life suddenly found the proposition much easier.

An increasingly influential French astronomer, Eugène Antoniadi, had come to doubt the existence of the canals after initially accepting the work of Schiaparelli and Flammarion as persuasive and correct.[77] During the 1909 opposition, Antoniadi observed Mars at the celebrated thirty-three-inch Meudon Observatory telescope, the largest in Europe. Though he observed for only nine nights during a month-long stay in Paris, Antoniadi reported seeing Mars so clearly at times that the linear appearance of the canals dissolved into an intricate mess of smaller, irregular details: "the geometrical 'canal' network is an optical illusion; and in its place the great refractor shows myriads of marbled and chequered objective fields, which no artist could ever think of drawing."[78]

Despite this comment, Antoniadi—an accomplished draftsman himself—nonetheless attempted to represent the complex markings he had seen, producing an image that looked more like Green's 1877 sketches than anything that had been produced in the intervening three decades (see fig. 2.15). He sent five sketches to Lowell with a letter describing his perfect certainty that they represented an objective view of Mars' surface. Commenting that the clarity of his observations "had surpassed all my expectations," he wrote, "I thought I was losing my senses; and it was only after seeing all these details constantly for hours that I concluded there was no doubt whatever regarding their objective reality."[79] Though Lowell had cautioned Antoniadi in an earlier letter about the danger that a large telescope such as Meudon's might actually show less detail (by allowing excess light to overwhelm subtle features), Antoniadi reported, "the tremendous difficulty was not to *see* the detail, but accurately to *represent* it."[80] Reprising part of the 1877–78 discussion between Green and Schiaparelli, Antoniadi claimed legitimacy for his sketches by touting his artistic skills: "Here, my experience in drawing proved of immense assistance, as, after my excitement, at the bewildering

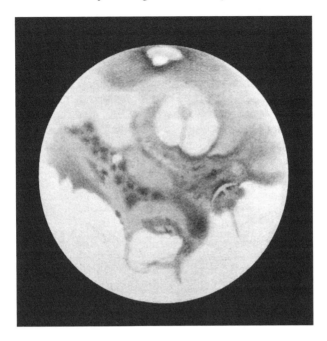

Figure 2.15. Published sketch of Mars by French astronomer Eugène Antoniadi, 1909. Antoniadi's depiction of Mars abandoned the abstract geometric style that had become dominant by the early twentieth century, returning to a more naturalistic style of shading such as that which had been used by English astronomer Nathaniel Green in 1877. This version was published as figure 99 in Maggini, *Il Pianete Marte*. Courtesy Richard McKim.

amount of detail visible, was over, I sat down and drew correctly, both with regard to form and intensity, all the markings visible."[81]

Lowell tried to discredit Antoniadi's claims, but to no avail. In personal letters, he suggested that Antoniadi's telescope aperture was so large it had caused a blurring effect. In response, Antoniadi only expressed even more certainty of what he had seen. He wrote to Lowell later in 1909, "I base all my ideas of Mars on what I saw myself at Meudon; and as I have not seen any geometrical canal network, I am inclined to consider it as an optical symbol of a more complex structure of the Martian deserts, whose appearance is quite irregular to my eye."[82] Antoniadi carefully and politely acknowledged that Lowell (and Schiaparelli) had discovered many real features on the Martian surface, but rejected the possibility that they could be anything other than natural.

Upon the occasion of Schiaparelli's death in 1910, Lowell wrote an eloquent obituary praising the Italian's canal discoveries while also blasting his own opponents for not accepting the reality of the canals.[83] It was to be, however, the last time he actively defended the inhabited-Mars hypothesis in a scientific publication, showing that the tide had finally turned. Antoniadi, on the other hand, wrote more than a dozen well-received scientific articles in 1909 and 1910, most of them directly refuting Lowell's theories. In his official reports for the British Astronomical Association, Antoniadi wrote with confidence and finality of the artificial canals' demise:

> We thus see in the so-called "canals" a work of Nature, not of Intellect; the spots relieving the gloom of a wilderness, and not the Titanic productions of supernatural beings. To account for their various phenomena, we need only invoke the natural agencies of vegetation, water, cloud, and inevitable differences of colour in a desert region.[84]

To understand how Antoniadi's nine nights of Mars observations succeeded in discrediting Lowell, who had a fifteen-year record of continuous observation and publication, we must consider the visual authority of Antoniadi's new claims in 1909. Upon completion of his stay at the Meudon Observatory, Antoniadi immediately began circulating his sketches to colleagues within the British astronomical community. At the same time, he wrote a series of articles about his and others' Mars observations in the *JBAA*. In most of these publications and letters, he emphasized the fact that his drawings showed more detail than Lowell's by revealing intricate detail in places where Lowell showed mere lines. He referred to a "vast and incredible amount of detail," claiming that "the fact that *no straight lines could be held steadily when much more delicate detail was continually visible* constitutes a fatal objection to their crumbling existence."[85]

Antoniadi and his ally, Maunder (an active Lowell critic), also pointed out that the new naturalistic, shaded sketches bore a striking resemblance to the latest photographs of Mars. Using the world's largest telescope (with a sixty-inch glass), the staff of the Mount Wilson Observatory in California had taken a series of photographs in 1909 that far exceeded Lowell's 1907 images in clarity and detail. Once again, however, the photographs failed to show any of the hard-edged features that commonly appeared in Lowell's drawings and maps.[86] Antoniadi's 1909 sketches thus appeared more accurate than Lowell's because of their similarity to the new photographic imagery.

Finally, it must be noted that Antoniadi's personal authority as a longstanding canal-theory supporter made him an especially effective critic. Antoniadi himself had reported seeing canals on numerous occasions and had drawn dozens of them on maps he compiled for the British Astronomical

Association in his capacity as the Mars Section director since 1896.[87] Furthermore, Antoniadi had championed the evidentiary quality of Lowell's 1905 and 1907 photographs. In an analysis published for the Royal Astronomical Society in 1908, for instance, Antoniadi commented that "the amount of detail shown on [Lowell's] photographs is very considerable" and noted that he could count seventeen canals as "more or less discernible on the images." Antoniadi thus could not be dismissed as a feeble observer who rejected the canals because he could not see them himself. He also shrewdly referred to other observers who had reported seeing irregular details within the canals in the last two decades, further supporting his claim.[88]

In the end, Antoniadi won a nearly complete reversal of the 1877 verdict, as his subtle, naturalistic shading won substantial approval from the astronomical communities in Europe and North America, relegating Lowell's hard-edged Schiaparellian-style maps to a weakened status as "startling theories."[89] Lowell himself made a last-ditch effort to revive interest for his theories, focusing on the centers of opposition in Europe. At presentations in London, Paris, and Berlin, he reportedly "presented his views so convincingly that many of his most violent opponents began to doubt" yet failed to overturn the growing European suspicion that the canals were illusions.[90] Maunder claimed at a meeting of the British Astronomical Association that the canals had been irrevocably put to rest:

> There never was any real ground for supposing that in the markings observed upon Mars they had any evidence of artificial action. Had it not been a sensational idea which lent itself to sensational writing in the daily press he [Maunder] did not believe they would ever have heard of it. He considered it was all the better for science that the idea was now completely disposed of. They need not occupy their minds with the idea that there were miraculous engineers at work on Mars, and they might sleep quietly in their beds without fear of invasion by the Martians after the fashion that Mr. H.G. Wells had so vividly described.[91]

Although his pronouncement was somewhat overstated, given that scientists were still seriously considering the possibility of Martian canals and intelligent life in the 1920s, Maunder accurately recorded a definitive reversal in scientific considerations of the geography of Mars.[92]

The reasons for this reversal include both photography's rise as a standard of proof as well as Antoniadi's claim that his few sketches showed more detail than Lowell's many maps. Where Green had argued in 1877 only that he saw something different than Schiaparelli, Antoniadi argued that he actually saw more than Lowell. Visually supported by the photographs—the new scientific imagery of truth—Antoniadi's sketches thus trumped Lowell's maps. After a long assault on the logic of Lowell's theory and the certainty of his methods, it was the dismantling of his map that finally diminished the scientific community's willingness to seriously entertain further talk of Mars' intelligent, canal-engineering inhabitants.

Conclusions

There is little value in assessing which early maps were "right" or "wrong" in terms of their faithfulness to modern-day imagery of the Martian surface. Maps produced at the turn of the twentieth century are much more valuable for what they reveal about the processes that served to authorize and legitimize certain views of the geography of Mars. The sharp rise of the inhabited-Mars theory in the late nineteenth century was intimately tied to the perceived objectivity of scientific cartography, the visual authority of specific maps, and the professional authority of various mapmakers. In essence, the apparent objectivity of cartography tended to conceal varying production processes, meaning that maps of Mars were compared and assessed primarily on the basis of their visual appearances, regardless of how they had been made.

As these maps gained self-referential power and iconic status on the basis of their detailed appearances, they also contributed to widespread popular enthusiasm for Lowell's inhabited-Mars theory. Because the inhabited-Mars theory was so keenly linked with the visual authority of the map and the privileged status of the most active Mars mapmakers, however, it was delicately dependent on the map's legitimacy. When the perceived objectivity of cartography faltered in the early 1900s in comparison with new photographic technologies, belief in Mars' supposed inhabitants lost considerable ground as well. The maps' waning credibility further weakened the position of astronomers like Lowell, whose stature as advocates of the inhabited-Mars theory was built on the foundation of their maps. By 1910, the astronomical communities of Europe and North America had largely abandoned their thirty-year flirtation with the idea of intelligent Martian inhabitants, returning to a naturalistic mapping (or sketching) style that closely resembled the pre-1877 maps and duplicated the appearance of Mars recorded in photographs.

Through maps, Mars became a geographical place, a contestable territory, and a celebrated locale for extraterrestrial life. In the same way that scientific maps allowed the British to conceptually subjugate India, the

French to justify their invasion of Egypt, and European explorers to depict an empty landscape in the heart of aboriginal Australia, maps of Mars authorized a new view of the red planet's landscape—as familiar, inhabited, and advanced.[93] Even though the maps eventually surrendered their authority, the underlying idea of Mars as an Earthlike and habitable world persisted. New explanations of the complex and detailed planet shown in Antoniadi's sketches and in the Mount Wilson photographs invoked natural life processes, even if they did not invoke intelligent Martians.[94] Cartography was thus integral to the origin and development of the scientific conceptualization of Mars as a world possibly inhabited. As Lowell himself put it, "Half the delight of travel consists in the pleasure of poring over maps in advance. [T]he charm of the chart grows all the greater in the case of a world which, in person, we have no hope of ever reaching."[95]

Representing Scientific Sites: Vision and Fieldwork at the Mountain Observatories

The altitude, the unusual steadiness, clearness, and dryness of the air at Flagstaff, render the Lowell equatorial by far the most powerful instrument in existence for the investigation of planetary detail. Only those who have studied Mars with this instrument, or the Amherst telescope, when at the recent opposition, it was mounted in South America, are competent to say what can be seen on Mars under the most favorable conditions.

—American zoologist George Agassiz (1908)

The data underlying the powerful Martian maps discussed in the previous chapter were produced in specific geographical locations, as were the maps themselves. The British astronomer Nathaniel Green lived and worked in a suburb of London but traveled to the Portuguese island of Madeira to record his most critical data on Mars. The Italian astronomer Giovanni Schiaparelli made his many observations of the Martian canals from a location in urban Milan that enjoyed a favorably southern latitude. The American astronomer Percival Lowell crafted his maps at a remote observatory in Arizona but later sent a photographer to the Chilean Andes in search of unimpeachable data. The French astronomer Eugène Antoniadi managed to impeach that very data with observations he conducted from suburban Paris. These places mattered, and not just in terms of what could or could not be seen at different latitudes and altitudes. They also weighed heavily on individual scientists' legitimacy as interpreters of Mars.

As outlined in the introductory chapter, recent scholarship in the history of science reminds us that science is never the placeless activity its practitioners so often hold up as an ideal. Examination of the "geography of science" reveals a spatial unevenness in the distribution of and access to facilities, instruments, scientific objects, allies, publications, and audiences. This unevenness definitively conditions the legitimacy of various scientists and ideas, helping to explain how and why certain ideas or theories come to be accepted as truth.

In examining the sites in which Mars science was produced and legitimized, we will see that the era of the Martian canal sensation coincided with a critical spatial transition in the discipline of astronomy. In the last decades of the nineteenth century and the early decades of the twentieth century, the locus of power in astronomical science began to move away from the low-altitude metropolitan observatories that had dominated the profession before the 1880s. Several of the most active participants in later phases of the canal debates conducted their observations at new high-altitude sites, particularly in the American West. Astronomers from the new Lick Observatory atop California's Mount Hamilton published regularly about Mars; Lowell emerged as the leading advocate of the inhabited-Mars theory from his mesa-top site near Flagstaff, Arizona; and numerous other astronomers traveled to the American West or to the mountains of South America to observe Mars and participate in the debate over its canals and supposed inhabitants.

These sites of Mars observation themselves quickly became important factors in the legitimization of Mars data. Numerous studies of Mars science and sensationalism have noted the important controversies that developed around the concept of "good seeing" as it related to observations of the red planet: over ideal telescope size, aperture settings, chromatic filters, and spectroscopic techniques. Beyond these technical issues, however, matters of location played a dominant role in the question of "seeing." As astronomy underwent several disciplinary transitions, the new American observatories mobilized a powerful representation of astronomy as a science that could be conducted properly only in mountain landscapes. The practice and representation of Mars science around the turn of the century reflected the power of this geographical association to influence both scientific and popular audiences. This chapter considers the links between physical location and scientific authority, showing how astronomers' site-specific acts of gathering data, interpreting findings, and making claims about Mars participated in disciplinary struggle at the turn of the century.

The New Geography of Astronomy

In the nineteenth century, a number of new observatories were built in remote and high-altitude sites, particularly in the American West. The geographical placement of these facilities was critical to their high-profile status and must be understood in the context of fundamental changes then underway in the discipline of astronomy. As astronomical investigations had turned away from the mechanical movements of celestial bodies and toward their composition and nature, astronomers found themselves performing observations and measurements that required more sophisticated equipment and more highly trained support personnel than ever before.[1] This in turn required significant financial investment, which was typically sought in the business community via philanthropists. The new observatories of the late nineteenth and early twentieth centuries were therefore large, expensive facilities that operated on a capitalist industrial management model aimed at maximizing the scientific return on benefactors' investments in telescopes and specialist staff.[2]

This trend, which was generally taking hold across the hard sciences as a whole, had a dramatic geographical effect on the practice of astronomical science. No longer content to rely on data gathered by legions of small telescopes in the backyards of dozens of amateurs, the new astronomy instead concentrated its scientific equipment and personnel in limited numbers of physical locations that could be controlled for maximum efficiency. These new locations themselves were carefully chosen to ensure that observatories could be in nearly continuous operation. To a shrewd and entrepreneurial-minded observatory director, it simply did not make sense to spend days or weeks waiting for good observing weather while expensive equipment and personnel sat idle. Nor would the new astronomy tolerate compromised legitimacy or certainty of research results on the basis of atmospheric impurity, as happened as a matter of course in nineteenth-century astronomy. New observatories had to be located in climatic and/or topographical zones likely to enjoy consistently clear skies and calm air.

Planetary observation, which did not require the same levels of highly specialized telescopes or expert professional staff as stellar observation, was caught up in the geographical transition nonetheless. A writer for the British journal *The Observatory* commented in 1882, for instance, that British skepticism over Italian astronomer Schiaparelli's reported Martian canals might stem from discrepancies of atmospheric conditions at different locations: "It is, of course, conceivable that markings which appear distinct and well-defined . . . when examined in the pellucid air of Northern Italy, would, in our unfortunate climate, be confused together, so as to give the appearance of faintly shaded districts."[3] Variations in observers' claims about the Martian surface features, in fact, prompted an attempt to standardize a numerical "scale of seeing" that could be used by all astronomers, regardless of location or instrument size. The standard scale was proposed as a way of addressing and resolving conflicts that occurred when multiple astronomers all reported "very good" seeing yet showed widely divergent findings in their sketches and maps. If all Mars astronomers were forced to calibrate their individual seeing scales (by analyzing the detailed appearance of diffraction rings around bright stars), it was suggested, "the excellence of any region in the most delicate astronomical work will thus be revealed with absolute impartiality."[4] Whereas "seeing" had previously been considered an atmospheric characteristic that varied from night to night in a given location, a fundamental reconception cast it as varying from location to location on a given night. Rather than fine-tuning one's instrumentation or method to cope with a certain location's seeing, it became preferable to change one's location in pursuit of better atmospheric conditions.

To the new astronomy, place thus mattered at the end of the nineteenth century in a way it had never mattered before, and the only places considered worthy of investment were at high altitude. Elevation above sea level had long been recognized as an important variable in astronomical seeing, given that atmospheric density (and the associated refraction or scattering of incoming light) decreases with height. As early as 1719, for example, Isaac Newton had remarked that the negative effects of atmospheric disturbance and density might be reduced by placing telescopes in locations of "serene and quiet air, such as may perhaps be found on the tops of the highest mountains above the grosser clouds." A few eighteenth-century expeditions to mountain slopes and summits had offered positive confirmation of this idea, followed by a series of conclusive tests linking mountaintop locations with atmospheric clarity in the 1850s.[5] Despite the inconveniences of situating cutting-edge scientific facilities in remote locations with difficult access, the benefits of high altitude were considered worth the effort. The world's first two major mountaintop observatories—on Sicily's Mount Etna (Bellini Observatory) and on California's Mount Hamilton (Lick Observatory)—opened their doors in the 1880s to widespread acclaim and anticipation.

By the last two decades of the nineteenth century, high-altitude location had come to be seen as such an important variable in astronomical work that elevation was usually one of the first things mentioned about an observatory. In an interesting twist, however, the perceived benefits of high-altitude astronomy came to be associated almost exclusively with mountain landscapes. Though a number of leading astronomers were of the opinion that high-altitude mesas or plateaus were more desirable than mountains for their benefits both of transparency (minimal refraction and scattering of light) and of tranquility (the absence of turbulent temperature gradients), popular astronomy publications focused attention on mountain topography to such an extent that the legitimacy conveyed by altitude seemed to apply only to those observatories in the mountains.

Lick Observatory's director, Edward Holden, stationed on a mountaintop himself, found the widespread popular association of mountain landscapes with good astronomy to be an opportunity, yet also a challenge. On the one hand, popular perceptions of mountains as sites of legitimate astronomical science contributed massively to his own observatory's undisputed prestige. On the other hand, the automatic association of mountains with good astronomy glossed over important issues of atmospheric variation in different mountain sites, making it necessary for Lick Observatory to claim a special status among other mountaintop observatories. In his 1896 *Mountain Observatories in America and Europe*, Holden tried to grapple with the complex link that had developed between observatory legitimacy and mountain topography. His monograph, published by the Smithsonian Institution, purported to be little more than a historical and geographical catalog of "the conditions of good vision at mountain stations all over the globe."[6] For the major observatories, Holden included climatic data and various analyses of the seeing conditions reported at the site. His book also included dozens of illustrations of mountain observatories, many of which were reprinted as plates in various volumes of *Publications of the Astronomical Society of the Pacific* around the same time the monograph was published (see, for example, fig. 3.1).

In the book, Holden delicately lamented the widespread belief "that *all* mountain-stations possessed striking advantages."[7] Subtly pointing out the flaws and problems of various other mountain observatories, Holden tried to craft a complex argument that all mountain locations were *not* created equal. He clarified for his readers that atmospheric transparency, which was achieved purely via altitude, was actually a secondary concern to atmospheric tranquility, which depended on many factors related to weather patterns, topography, and vegetation, among others. In this rhetoric, Holden intended primarily to assert the superiority of his own Lick Observatory over the many other mountaintop stations profiled in his book, for he noted again and again that the sky above Mount Hamilton consistently experienced marvelous transparency *and* tranquility.

Yet Holden's discussion of the distinction between transparency and tranquility led him to admit that the latter could actually be found more easily on plains than slopes. He allowed, in fact, that "[o]ther things being equal, an astronomical station on an extended and elevated plain is preferable to one on a sharp peak. The conditions for level and tranquil

Figure 3.1. Illustration of Nice Observatory, 1895. This plate emphasizing the scenic mountain location of the Nice Observatory appeared in Holden, *Mountain Observatories* (facing
p. 19). It had previously appeared in *Publications of the Astronomical Society of the Pacific*
as part of a series of plates showing European and American observatories.

arrangements of air-strata are more favorable in the former case."[8] In a book
ostensibly meant to catalog and celebrate the achievements of mountain observatories, this comment by the director of the world's pre-eminent mountaintop observatory is somewhat jarring. Holden's nod to the astronomical
suitability of level plains must be read as a desperate attempt to combat
what he considered a pervasive and thoughtless assumption that all mountain observatories were legitimate, regardless of their specific characteristics.
Holden's concerned comments show how entrenched the perceived link between mountains and scientific authority had become. Invisible conditions
like atmospheric transparency and tranquility—the reasons for locating at
high altitude—had much less impact on an observatory's credibility than
visible conditions like dramatic mountain topography. In the eyes of both
scientific and popular audiences, observatories located in or on mountains
were automatically assumed to be authoritative producers of legitimate astronomical knowledge.

Astronomy and the American West

Despite Holden's calculated caution, the professionalization and industrialization of the new astronomy would remain strongly linked to mountain

geography, especially in the United States. Several new observatories built in the American West around the turn of the twentieth century were hailed as great advances for the discipline and for American science because of their large telescopes, professional staffs, and commitment to new areas of stellar research. But the profile and status of these observatories were also highly dependent on their mountain locations. Those observatories in remote mountain locations were automatically taken seriously, even with smaller telescopes and less-experienced astronomers. Those closer to metropolitan areas and on the plains, in contrast, struggled to establish legitimacy, despite having large telescopes and highly trained staff astronomers.

The University of California's Lick Observatory, established in 1888, was the first "big-science" institution in the United States. Endowed by California businessman James Lick, the observatory was envisioned from the beginning as a world-class institution that would outshine all other observatories in two regards: it would have the most powerful telescope in the world, and it would be sited on a mountaintop with excellent seeing conditions.[9] Lick's predilection for a mountain site was influenced by the recent enthusiasm among astronomers for high-altitude sites, which he understood "would amply repay the inconvenience" of transporting materials and asking astronomers to live in difficult and isolated conditions.[10]

Lick himself was involved in the site selection and apparently gave his blessing to the remote peak of Mount Hamilton in California's Diablo range partly because he was enchanted by the fact that he could see its summit from his own home near the south end of San Francisco Bay.[11] At 4,200 feet high, the mountain exceeded Lick's minimum elevation criterion of 4,000 feet. This height and California's reputation for clear skies created a powerful assumption that the site would be ideal for astronomy. No formal evaluation of Mount Hamilton's atmospheric characteristics was performed, in fact, before the site had been officially selected and the County of Santa Clara had been induced to build an expensive twenty-six-mile road to the mountaintop site. (The road was built in 1876, but an official test of "seeing" conditions was not conducted until 1879.)[12] Given the size of Lick's investment and the height of the astronomical community's hopes for its new centerpiece observatory, it is perhaps fortunate that when the site was eventually tested, it was indeed determined to have very good atmospheric characteristics! Before this determination, however, Mount Hamilton had already received favor, not because of the particulars of its scientific advantages but more because it fit the generalized notion of the mountain as a proper location for astronomy. As the editor of an astronomical journal remarked before the observatory was even complete: "One year on the summit of the California mountains affords the opportunities which twelve years of observations in the changeable climates of other states do not furnish."[13]

Lick's second ambition—to have the world's most powerful telescope at his observatory—was also fulfilled, but this quality was rarely remarked without simultaneous reference to the observatory's mountain location. Holden participated in this emphasis, issuing a widely circulated informational pamphlet that visually emphasized the observatory site, rather than its famous telescope.[14] Of the pamphlet's fifteen images, four of the first six depicted the mountaintop site—as remote, overgrown, menacing, or sublime (see, for example, figs. 3.2 and 3.3). The pamphlet contained only one image, near the end, of the great equatorial telescope, then the largest in the world. The mountain location of the Lick Observatory was thus a major

LOOKING SOUTH-WEST TOWARDS MT. HAMILTON

Figure 3.2. Illustration of Lick Observatory site, 1895. This image, which was one of the first illustrations included in Holden's 1895 *Brief Account of the Lick Observatory* (on p.14), is typical in that it dramatized the remoteness of the facility's mountaintop site and showed no means of access.

THE ROAD TO MT. HAMILTON.

Figure 3.3. Mount Hamilton, 1895. This photograph, which was one of the first illustrations
included in Holden's 1895 *Brief Account of the Lick Observatory* (on p. 4),
portrayed the landscape of Mount Hamilton as somewhat difficult of access.

component of its status and credibility, quite separate from the observatory's actual work and contributions to research in stellar astronomy.

The second large-scale American observatory, the University of Chicago's Yerkes Observatory, provides another example of the role mountain geography played in establishing credibility for American observatories. Like the Lick Observatory, Yerkes was funded by a philanthropist who wanted his observatory to boast the largest telescope in the world. Much of the drama surrounding the new observatory's planning and construction in the 1890s, in fact, focused on its attempt to "lick the Lick" by installing a telescope with a forty-inch lens, which would famously exceed the thirty-six-inch lens of Mount Hamilton's celebrated instrument. The Yerkes Observatory was conceived as a centerpiece of the University of Chicago and of the city of Chicago, both then emerging on the national and international stages. In fact, the observatory's first director used the occasion of Chicago's coming-out

party—the 1893 World's Fair—to organize the first international astronomical congress ever held in the United States, even though the Yerkes
Observatory was still only in the planning stages at that time. The planned
facility featured prominently at the congress and showcased the promise of
American astronomy and of the University of Chicago.[15]

There was only one problem with the Yerkes Observatory: its location.
Given the University of Chicago's desire to maintain a close association
with one of its showcase units, a site was selected for the observatory in Lake
Geneva, Wisconsin, which was "then just at the limit of leisurely commut-
ing distance by train" from Chicago. It was also, coincidentally, a "resort for
the choicest people of Chicago," whom the University of Chicago president
wanted to lure as donors.[16] Although the site selectors had studied a number
of environmental and atmospheric factors, satisfying themselves that the
Lake Geneva location was well suited for astronomy, the observatory's
spatial association with the city and the easy life proved to be a constant
hindrance. Holden, of the rival Lick Observatory, for instance, boldly noted
the new observatory's near-urban location in this damning critique:

> [T]he site chosen for the 40-inch Yerkes refractor of the University of Chicago
> lies about midway between Chicago and Madison. Unless the conditions at
> Lake Geneva, Wisconsin, are distinctly better than those of the region near by,
> its selection as a site for the largest of telescopes may turn out to have been
> an error of judgment.[17]

Yerkes Observatory director George Hale found himself constantly defending the site-selection process and decision, thus revealing deep concerns about the site's influence on observatory legitimacy. He responded to
criticisms like Holden's in two ways. First, Hale provided technical explanations of the site's atmospheric suitability, publicizing the fact that he had
administered a questionnaire to many prominent astronomers concerning
the effects on astronomical research of proximity to urban areas, to lakes,
and to railroads.[18] He even printed their verbatim responses to the questionnaire, thus relying on the stature and credibility of others to support his
view that the Lake Geneva location posed no major detriment to the observations planned for the new observatory. Hale's second primary response
was to emphasize his observatory's remoteness as a way of perceptually dis-
tancing it from urban Chicago. At the observatory dedication in 1897, for
instance, he thanked attendees for traveling to a site so far "removed from
the neighborhood of great cities, and from the more populous regions of

the United States," though in fact most of them had taken only a short train-ride from Chicago.[19]

In these representations, Hale was forced to acknowledge the favoritism usually shown to mountain sites, particularly that of the Lick Observatory. In his transcription of prominent astronomers' responses to his questionnaire about ideal conditions for an observatory, he disclosed the following comment by Simon Newcomb (then considered the leading American astronomer): "To be of the greatest benefit to science the telescope should be mounted at some such point as Mt. Hamilton, California; Arequipa, Peru; or the Peak of Teneriffe." Although Newcomb had otherwise supported the Lake Geneva location, he clearly indicated that a mountain site would be preferable. Hale tried to rebuff such favoritism by arguing that mountain locations were not *necessarily* a guarantee of good astronomical research and by arguing that his Lake Geneva institution was blessed with very good atmospheric conditions. He tried in vain to convince readers that "notwithstanding a widespread impression to the contrary, the excellent atmospheric conditions enjoyed at the Lick Observatory do not seem to be common to all mountain summits."[20] He also drew attention to his colleague Edward Barnard's experience conducting nebula observations at both Yerkes and Lick, reporting that "Professor Barnard has found that the best nights here are fully as good as the best nights at the Lick Observatory . . . and he assures me that he now sees [certain nebulae] better than he could see them with the Lick telescope."[21]

Despite these efforts, however, Yerkes was persistently dogged by accusations of "bad seeing." Several of Yerkes' own astronomers admitted their site's inferiority to Mount Hamilton as a matter of fact. Sherburne Burnham, a respected double-star observer at Yerkes who had also performed the official atmospheric testing for Lick Observatory in 1879, wrote in 1900 that "there is probably no place in the world, where an observatory has been established, which can compare favorably with Mount Hamilton."[22] Hale himself soon grew tired of the difficulties of defending Yerkes. He left Chicago in 1903 to establish a new solar observatory outside Pasadena on Mount Wilson, elevation 5,700 feet.[23] Despite Yerkes Observatory's massive telescope, its generous funding, its meticulous organization, and its soaring expectations, it never managed to rise above concerns about its location. Yerkes was considered an excellent site by eastern or midwestern standards, but could not truly challenge the western mountain sites for prestige.

By contrast, the Lowell Observatory—more meagerly equipped, funded, and staffed than either the Lick or Yerkes observatories—managed to achieve considerable acclaim by promoting the excellent conditions of its site above Flagstaff, Arizona, elevation 7,000 feet. In an era of increasing spectral and stellar work, an observatory dedicated to the visual investigation of a single planet seemed an anachronism. As such, amateur astronomer Lowell could hardly have expected to earn much esteem among professional astronomers and major observatories with his research program focused solely on the planet Mars. He did not help his case by publishing observational findings alongside speculative interpretations of the Martian surface as an inhabited landscape. Lowell's propensity for taking quasi-scientific arguments directly to popular audiences through magazines and lectures seemed to go against every promising trend in American astronomy. In thus antagonizing leading American astronomers, Lowell inspired numerous assaults against his own and his observatory's legitimacy.

And yet Lowell somehow managed to establish and maintain significant credibility, especially in the public eye. One of the most important things he did in this regard was to emphasize the remoteness of his observatory's location, the superiority of its altitude, and the excellence of its climate. In his publications, he regularly emphasized that he had investigated climatic conditions in numerous western sites before selecting high-altitude Flagstaff "for the purpose of getting as good air as practicable."[24] He relied on this fact heavily in asserting that his observatory was much more credible than any on the East Coast or in urban areas, lamenting that "at the present time most observatories are situated where man is greatly handicapped in his own efforts toward the stars" by city smoke, electric lights and other pollutants of atmospheric visibility.[25] He even went so far as to argue that his observatory was on equal footing with the world-class Lick Observatory by virtue of his advantageous location, despite the great difference between their telescope powers and staff experience.

Popular writers and audiences responded enthusiastically to this strategy, regularly commenting on the advantages of Lowell Observatory's high-altitude location when discussing the Mars debates. Though professional astronomers generally did not express any enthusiasm for Lowell's theory-driven methods, his speculative hypothesis, and his targeting of popular publications, they often found themselves forced to admit the quality of his location. Leading American astronomer Simon Newcomb, who never accepted Lowell's theory about the inhabitants of Mars, nonetheless wrote of the Flagstaff observatory that "its situation is believed to be one of the best as regards atmospheric conditions." Lowell encouraged such comments with his own highly publicized attempts to find a site better than Flagstaff.

He investigated a site in northern Mexico in 1896, eventually determining that Flagstaff was still superior. He also traveled to Algeria to investigate possible sites, drawing this comment from the British publication The *Observatory*: "He is looking out for the best climate he can get. Notwithstanding he is at present very well satisfied with his position at Flagstaff, Arizona; and his account of the conditions there is certainly enough to fill one with envy." Such comments indicate the extent to which geographical location had achieved parity with other factors, such as professional rank and power of instrument, which also defined an astronomer's credibility.[26]

In the context of this chapter, it must be noted that Lowell Observatory is not in the mountains. Rather, it is located on a high mesa. This fact apparently escaped many of Lowell's readers and audiences at the time, however, as his observatory was frequently assumed to be in the mountains by virtue of its reported remoteness, altitude, climate, and general location in

Figure 3.4. Flagstaff's San Francisco Mountains, 1906. This photograph was included in the introduction for Percival Lowell's most widely circulated book on Mars, *Mars and Its Canals* (on p.18). Although the Lowell Observatory was not located in the mountains shown, the peaks were visible from the observatory's Flagstaff, Arizona, location. Lowell referred to these mountains and emphasized their proximity (approximately twelve miles) both textually and visually in numerous publications.

Figure 3.5. Lowell Observatory telescope dome, 1906. This photograph
was included as the frontispiece for Percival Lowell's most widely circulated
book on Mars, *Mars and its Canals*. It depicts the observatory's main telescope
dome in a wilderness setting, visually prioritizing the ponderosa pine forest that
dominated the mesa on which the observatory had been constructed.

"the West." The well-known French astronomer Flammarion, for example,
referred in publication to the excellent climate of the "Arizona Mountains"
and lauded Lowell's site on "Flagstaff Mountain," which does not actually exist.[27] Lowell and his associates did nothing to correct this frequent
mistake. On the contrary, they actively cultivated a close representational

association with mountains. One of Lowell's small staff described the observatory's location thus: "It is a trifle short of 7,000 feet above the sea and is ten miles south of the San Francisco Peaks whose highest point is 12,800 feet in elevation."[28] The San Francisco Peaks so prominently noted in this quote had nothing whatsoever to do with the observatory, but they (and their height) were regularly mentioned in connection with the observatory. Lowell's first book, *Mars*, actually included photographs of the San Francisco Peaks (fig. 3.4) alongside photographs of the observatory buildings (e.g., fig. 3.5), implying that the observatory was in fact in the mountains. Both Lowell and his most experienced staff astronomer, William Pickering, were members of the Appalachian Mountain Club and were known as enthusiastic recreational climbers, which deepened the observatory's connection to mountain landscapes.[29]

As far as the public was concerned, the Lowell Observatory was a "mountain observatory" just like the Lick Observatory. Its perceived links to dramatic mountain topography gave it an enviable level of credibility. Astronomers who knew better, however, were nonetheless impressed by Lowell Observatory's location. The high-altitude mesa site enjoyed the advantages of atmospheric transparency yet suffered none of the disadvantages of atmospheric turbulence that plagued many mountain sites. It thus had the best of both worlds: a high-altitude nonmountain site that achieved widespread popular acclaim on the basis of erroneous assumptions that it *was* in the mountains. The first condition lent Lowell Observatory credibility in the eyes of other astronomers. The second helped it cultivate popular legitimacy.

A Pure View of Mars from the Mountains

The question remains as to why exactly astronomical observatories achieved such unassailable credibility in the public eye through an association with mountain landscapes. Why are mountains so important? At a very basic level, the maneuverings of astronomical scientists reflected a cultural shift that had begun to imbue mountains with a new and critical importance. As mountains came to be seen in art and literature as places of transcendence and divinity, they also came to be see in science as sites of purity and vision. This section traces the way these cultural ideals influenced Mars astronomy, focusing specifically on several key claims regarding scientific vision. Those astronomers and observatories who established legitimacy and authority most successfully used mountain representations to assert that their vision was pure—in environmental, visual, and moral terms.

Starting in the eighteenth century, prevailing European notions of mountains as frightening or dangerous places began to give way to a new view of mountains as sublime locations of divinity and human transcendence. The geometric fact that mountains offered a panoramic and enlarged perspective came to be celebrated by Romantic artists and authors who glorified the individualism of the summit-achieving climber as well as the ideal of liberty symbolized by the open mountaintop.[30] As the pursuit of a sublime view from altitude became a staple of both travel and artwork in Europe and North America, "the mountain-top and the viewpoint became accepted sites of contemplation and creativity: places where you were brought to see further both physically and metaphysically."[31] By the middle of the nineteenth century, altitude was also thought to have a cleansing and purifying affect on the human body and mind.

Mars astronomers reflected the shifting cultural importance of mountains in their assertions of legitimacy and in their characterizations of astronomy as pure, untainted, and objective. Most often, mountains were said to provide pure vision in environmental terms. With increasing altitude, air becomes less dense and contains fewer particles, meaning there are fewer opportunities for air molecules to impede or refract the path of light as it passes through Earth's atmosphere. All other things being equal, distant celestial objects thus appear brighter from high-altitude positions than they would from sea level. Furthermore, high-altitude sites provide the opportunity to rise above dense cloud cover and escape the visual distortion caused by water-vapor molecules. Although this advantage is less dramatic in regions dominated by dry air, it was certainly a major consideration for astronomers working in humid regions. Finally, many of the mountain observatory sites were in rather remote locations, allowing them to boast a freedom from light pollution and urban smog. Taken together, these environmental characteristics were reported to provide a remarkable atmospheric purity that rendered mountains sublime sites for astronomical science.

Both the Lick Observatory and the Lowell Observatory relied heavily on representations of environmental purity in their claims about Mars. Lick was the first American observatory to report seeing the Martian canals, and director Holden chalked up his observatory's 1888 confirmation of European canal sightings to the pure atmosphere above Mount Hamilton. In fact, Holden went so far as to dispute various European observers' findings about the nature of the canals that year, even though it was his observatory's first experience with Mars observation. He felt justified in making these critiques because of Lick's sublime location, which, he claimed, allowed for certain

Martian features to be seen distinctly, even in suboptimal conditions.[32] As popular interest in Mars began its dramatic rise, Holden continued to focus on Lick's environmental purity as a way of asserting credibility for the observatory. Though Mars was very low in the northern sky during the 1892 opposition, Holden reported that the sublime conditions at Lick allowed for numerous observations and sketches of Mars at a time when most other American observatories reported a dismal failure in their attempts to get good views of the red planet.[33]

The perfection of Mount Hamilton's location was again a theme in 1894 when Lick astronomer Wallace Campbell tackled the conventional wisdom about water vapor on Mars. Several prior studies had reported evidence that the Martian atmosphere contained water vapor. In 1894, however, Campbell published controversial spectrographic findings that showed little or no water vapor on the red planet, thus provoking a strong reaction among the inhabited-Mars proponents, including Lowell. In making his case, Campbell referred to the "extremely unfavorable" atmospheric conditions under which past observations had been made, referring to both the altitude and relative humidity of past observations. He then lauded the high altitude of Lick Observatory, "which eliminates from the problem the absorptive effect of the lower 4,200 feet of the Earth's atmosphere, with all its impurities. Most of the old observations were made from near sea-level."[34] Campbell thus focused on the environmental purity of the Lick Observatory mountaintop site to validate his controversial position regarding the science of Mars.

Percival Lowell engaged in similar representational maneuvers in his own publications, especially when trying to demonstrate that his new observatory was a legitimate site of scientific knowledge-making. Lowell regularly included images of mountains in representations of his observatory. He also opened nearly every publication about Mars with a discussion of the clarity of his observatory's high-altitude views. Although Lowell was a newcomer to professional astronomy, he successfully repelled attacks by Yerkes Observatory's Hale, the American Nautical Almanac's Newcomb, and a number of British skeptics by focusing on the issue of environmental purity. Essentially, he turned attention away from the content of his detractors' critiques and toward the location of their urban observatories. Hale and Newcomb, of Chicago and Washington, D.C., respectively, both suffered Lowell's sarcasm regarding their eastern, near-urban locations. According to Lowell, no one was qualified to critique his research unless he or she was working in similar or better atmospheric conditions. He found it especially easy to attack his British critics, given that British astronomers themselves were frequently given to lamentation about the atmospheric impurities of the British Isles. The British publication *The Observatory*, for example, allowed in an editorial comment that British astronomers' inability to see Schiaparelli's canals might be due to the difficulties of observation in "our unfortunate climate."[35] Lowell suggested in a letter to Walter Maunder of the Royal Observatory, Greenwich that "if England would only send out an expedition to steady air . . . it would soon convince itself of these realities [the canals]."[36] In this rhetoric, an inability to see the Martian canals was linked to impure or polluted observing sites. His own remarkable ability to see increasing numbers of canals, on the other hand, could be chalked up to the environmental purity and sublimity of his site.

A second way that western astronomers used the idea of mountain purity to cultivate legitimacy was by discussing the geometric perspective available from mountain locations. Although astronomers' scientific claims ostensibly depended on their upward views into the night sky, many of them commented on the sublime *downward* vistas their observatories offered of surrounding terrestrial landscapes. From a mountaintop or mesa cliff, the astronomer's view of his home planet was said to be spectacular. When the Lick Observatory was opened, for example, *Scientific American* reported enthusiastically on the new facility's view of California:

> Professor Holden says: 'It would be difficult to find in the whole world a more magnificent view than can be had from the summit just before sunrise, on one of our August mornings. The eastern sky is saffron and gold, with just a few thin horizontal bars of purple and rosy clouds. . . . The instant the sunbeams touch the horizon the whole panorama of the Sierra Nevada flashes out, 180 miles distant. . . . The Bay of San Francisco looks like a piece of a child's dissecting map, and is lost in the fogs near the city. The buildings of the city seem strangely placed in the midst of all the quiet beauty and the wild strength of the mountains. Then you catch a glimpse of the Pacific in the southwest and of countless minor ranges of mountains and hills that are scattered toward every point of the compass, while, if the atmosphere is especially clear, you can plainly see to the north Mount Shasta, 175 miles distant.[37]

Not only did this detailed representation garner attention from popular audiences who were interested in the landscapes of the American West, but it also conferred authority on all vision claims coming from the Lick Observatory. This representation implied that if Holden could see with such clarity beyond the fogs of San Francisco, all the way to majestic Mount Shasta, then Lick Observatory's claims for seeing the surface of Mars must surely be trustworthy as well.

Likewise, Percival Lowell used his high-altitude perspective on Arizona as a means of cultivating legitimacy for himself and his claims about Mars. Going a step further than Holden's celebration of the beauty and sublimity of his terrestrial perspective, Lowell argued that working in a high-altitude, desert environment gave him a critical interpretive perspective on the red planet. Aside from their atmospheric purity, Lowell argued, high deserts provided clues to what the surface of Mars was probably like. "Though partial only, the features and traits of our arid zones are sufficiently like what prevails on Mars to make them in some sort exponent of physical conditions and actions there." Because Lowell believed that Earth's desert belts were "the beginning of that stage of world evolution into which Mars is already well advanced . . . symptomatic of the passing of a terraqueous globe into a purely terrestrial one," he believed his observatory's location in Arizona gave him a unique perspective on and understanding of Mars.[38]

Additionally, Lowell argued that working at high altitude helped him understand what forms life on Mars might take. He frequently used the nearby San Francisco Peaks as his example of a terrestrial environment in which plants and animals thrived despite very low barometric pressures and very cold average temperatures. As he wrote in one of his major books on Mars: "On the high plateau of northern Arizona and on the still higher volcanic cones that rise from them as a base into now disintegrating peaks, the thin cold air proves no bar to life."[39] This comment was meant to directly rebut one of the primary objections to his theory of Martian life, which held that the planet's cold temperatures and thin atmosphere would preclude the possibility of any inhabitants. It also managed to maintain a representational link between Lowell Observatory and the San Francisco Mountains, which was so important to the popular perception that Lowell's was a mountain observatory.

The third way in which astronomers used their mountain locations to represent their work as "pure" and legitimate centered on a supposed link between personal isolation and clear vision. Because of the remoteness of their high-altitude and mountain sites, western astronomers represented themselves and their scientific claims as morally pure and trustworthy. As was shown in the previous section, official discussion of the Yerkes Observatory site during its construction and opening had included many derogatory comments about the potentially polluting influence of Chicago. (Lowell once remarked to an audience of British astronomers that "the Yerkes air was 'born bad.'"[40]) But western astronomers represented the superiority of their locations as much deeper than the mere fact of their geometric distance from heat, light, and smog; they represented cities as morally barren sites where pure vision was impossible. Lowell was perhaps most eloquent and persuasive in arguing that high-altitude astronomers were free from corrupting influences that would otherwise denigrate the purity of their investigations, observations, and intentions. In his representations, proper investigations of Mars could be done only in high, remote places:

> [The astronomer] must abandon cities and forego plains. Only in places raised above and aloof from men can he profitably pursue his search, places where nature never meant him to dwell and admonishes him of the fact by sundry hints of a more or less distressing character. . . . Withdrawn from contact with his kind, he is by that much raised above human prejudice and limitation. [41]

In disagreements about the nature of Mars, western and mountain-based astronomers easily invoked this characterization of their urban-based opponents as fundamentally unable to see. William Pickering used it bluntly, for instance, to reject doubts about the existence of the canals: "An astronomer who has never looked through a telescope, except in northern Europe or the eastern United States, has no right to express any opinion on the subject, because he simply does not know what good seeing looks like, and his opinion is therefore valueless. He might as well express his views on electro-dynamics or physiology."[42]

In the American context, Lowell and other western astronomers were tapping into a national enthusiasm for wilderness that was just then emerging in the United States.[43] This enthusiasm was connected with the Romantic idea that wilderness was a locus of purity and clarity, where human senses were most finely attuned and humanity's perception of sublime nature was most direct. As the new American nation had come to define its character and culture in terms of a unique engagement with wilderness and frontier landscapes, the nineteenth-century closing of the frontier and the rapid loss of wilderness became a national concern. America's first champion for the preservation of wilderness, John Muir, focused his attention heavily on the western mountains, casting them as sublime antidotes to the dirty and corrupt (eastern) cities that had not yet cast off their European character to become fully American. At the same time that these mountains became potent symbols of the American nation and character, astronomers' representations of their scientific exploits in the sublime air of the American High West gained great authority with general audiences. In this cultural climate, Lowell found it easy to generalize about the inferiority of the East for astronomy: "Not till we pass beyond the Missouri do the stars shine out as they shone before the white man came."[44] Although the charge that metropolitan observatories and astronomers were necessarily laboring with impaired vision (and a character deficiency) certainly did not sit well with the many American and especially British astronomers who were working in urban sea-level environments, it was effective. Western mountain landscapes were effectively invoked as morally, visually, and environmentally pure sites of uniquely legitimate Mars science.

The Wilderness Challenge

Alongside these representations of the sublime beauty and purity of mountain sites, astronomers also emphasized the ruggedness and physical challenge associated with working in the mountains. Textual descriptions and graphical depictions of snow, ice, bad weather, and dangerous terrain reinforced the concept that the best investigations of Mars were being done in difficult, dangerous wilderness settings. In such representations, the astronomer was painted as a masculine individual, confronting mountain wilderness and rising to its challenges in the name of science. According to western astronomers' claims, knowledge of specific landscapes was critical and the ability to conduct oneself properly within a challenging environment paramount. This section explores the ways that astronomers cultivated legitimacy for themselves (and challenged others' legitimacy claims) on the basis of their ability to perform difficult physical operations within mountain-wilderness field sites. This kind of representation worked hand in hand with the representations of mountains as sublime scientific sites, producing a complex geography of astronomical legitimacy throughout the debates over Mars.

Starting with the first mountain observatories, astronomers invariably described their work locations in terms of rugged environmental conditions. For all the discussion of Mount Hamilton's sublime location and perfect seeing, those who worked there regularly took pains to point out its difficulties. Perhaps the most common theme concerned the extreme discomfort of mountaintop winters. Quoting an address by Holden to the Astronomical Society of the Pacific, *Scientific American* reported breathlessly on the harsh winters astronomers faced at the new observatory:

Three or four times during each of the winter months the wind blows at the rate of more than sixty miles an hour. Many of the stronger gusts, which must exceeded seventy-five or eighty miles an hour, have never yet been measured, for no instrument can be found that will stand the test. Although the windmill which supplies the observatory with water is carefully furled before each storm and held in position by iron braces, nearly two inches in diameter, once a year it is torn from its mounting and destroyed.

During five days of February, 1890, absolutely no communication with the outside world was possible. The snow fell in enormous quantities, and a fierce blizzard was blowing, which could not be faced.[45]

Even indoors, the article continued, astronomers suffered greatly from extreme cold and a lack of firewood and supplies that would normally be brought to the peak by stagecoach. When the snow was too deep for deliveries, parties of astronomers were forced to hike to the base of the mountain to pick up supplies and bodily carry them back to the summit. Under these conditions, neither the observatory rooms nor the residences could be made comfortable, with Holden reporting that "water froze on the very dining table" during the severe winters of 1886 and 1887.[46] This is one of numerous comments meant to convey the extreme discomfort of life on Mount Hamilton in the winter. Based on the challenging characteristics of the mountaintop site, astronomers could be portrayed as heroes who cheerfully accepted extra labor and persevered with their work under extremely harsh conditions.

A related theme in the representations of Lick Observatory focused on the natural hazards associated with working on Mount Hamilton. Lick astronomers regularly published statistics on the number and intensity of fires and earthquakes they observed, giving the impression that California in general and Mount Hamilton in particular were constantly facing grave natural threats.[47] Although such hazards ostensibly had an impact on astronomical seeing conditions, the prominence of their representation in astronomical journals shows that they also served to emphasize for readers the dangers and difficulties of working in a mountain-wilderness setting. The objective statistical nature of earthquake and fire observation reports did not diminish their effectiveness in associating the observatory with natural hazards. When *Publications of the Astronomical Society of the Pacific* published its series of images showing mountain observatories between 1893 and 1896, for instance, it included a dramatic image of smoke rising from a forest fire near Mount Hamilton (fig. 3.6). The image did not actually show the Lick Observatory itself, but the implication was clear: the observatory was largely characterized by its proximity to natural hazards.

FOREST FIRE NEAR MOUNT HAMILTON, AUGUST, 1891
FROM A NEGATIVE BY PROFESSOR CAMPBELL.

Figure 3.6. Forest fire near Mount Hamilton, 1891. This plate was one of many illustrations of mountain observatories published in *Publications of the Astronomical Society of the Pacific* between 1893 and 1896. It appeared as the frontispiece for the March 1893 number, showing smoke rising from a fire near Mount Hamilton two years before the photograph's publication. The image emphasized astronomers' proximity to the blaze by noting that the photograph was taken by one of the Lick astronomers, W. W. Campbell.

Representations of the Lick Observatory and environs also typically emphasized the facility's difficult remoteness. Graphically, remoteness was conveyed in images that omitted a view of the road (refer to fig. 3.2 above). In images where the road *was* shown (as in fig. 3.3 above), it typically appeared overgrown or difficult to access. Astronomers and visitors were always shown ascending the road on foot or horseback, whereas textual records indicate that stagecoach and wagon were the most common ways of reaching the summit. More generally, *Scientific American* noted that the observatory was twenty-six miles from the nearest settled area, San Jose, then quoted Holden's frustration regarding the delays in getting scientific equipment to such a remote site. "For example, a bit of colored glass is wanted to moderate the brightness of Mars, so that the satellites can be seen. Where is it to be found? There is not so much as a square millimeter of such glass west of the Allegheny Mountains."[48]

The powerful implication of these seemingly innocent representations was that it took a special kind of scientist to endure the discomforts, hazards, and isolation of Mount Hamilton. Despite these difficulties, *Scientific American* concluded, the Lick Observatory continued to do superior astronomical work, "as the large number of observations in every periodical proves."[49] What Holden himself may or may not have realized is that his observatory achieved success not *despite* the difficulties of working in a remote mountain site, but in some measure *because of* the site's legitimizing effect on Lick's authority.

The debates over Mars, its canals, and its inhabitants trafficked heavily in representations of mountain observatories as challenging wilderness sites. Lowell, especially, referred regularly to the ruggedness of his observatory's environment when discussing his scientific findings about Mars.

> Like mundane exploration, it [astronomy] is arduous too; *ad astra per aspera* is here literally true. For it is a journey not devoid of hardship and discomfort by the way. Its starting-point precludes as much. To get conditions proper for his work the explorer must forego the haunts of men and even those terrestrial spots found by them most habitable. Astronomy now demands bodily abstraction of its devotee.[50]

In other words, good astronomy can be done only when the astronomer is physically present in the wilderness, where attendant discomforts play a critical role in the exploration process. No matter that the Lowell Observatory was actually within easy walking distance of downtown Flagstaff and its railroad. Lowell perceived and represented his location as a frontier outpost, where hardy men were challenged by their difficult environment to develop special skills that were not available to others.

In many cases, this representation was evoked implicitly through counterportraits of urban or sea-level sites (and astronomers) as untrustworthy or "handicapped" in terms of their Mars claims.[51] One of Lowell's staff astronomers remarked after a visit to the East Coast that Flagstaff's perception of Mars "was so vastly superior to the best that was visible in Washington that the seeing seemed to be not merely improved quantitatively but to be of an entirely different quality."[52] Part of the difference in quality had to do with the fortitude of astronomers who endured the difficulties of living and working at high-altitude facilities. In his publications and correspondence, Lowell developed this theme by rhetorically challenging his critics to visit the Lowell Observatory, as in this letter to a British astronomer: "Since 'seeing is believing' I have an idea. Why will you not come out to Flagstaff this autumn and observe the canals of Mars for yourself? I make no doubt that with a little practice in this class of observations you will be able to see them perfectly well."[53] In such challenges, Lowell implied that only those astronomers who were hardy enough to undertake a westward journey were capable of good scientific vision.

In addition, Lowell made it clear that he believed the wilderness conditions of a mountain observatory site honed the skills of the astronomer. He often referred, as above, to the "practice" or "training" required before an astronomer could see Mars properly from a good mountain site. According to Lowell, it took time to adapt not only to better seeing conditions but also to the new environment. A zoologist associate of Lowell's echoed this idea after a six-week visit to Flagstaff during which, he wrote, he had gradually come to see Mars more and more clearly as he adjusted to the conditions and trained his eye.[54] Another nonastronomer associate of Lowell's argued after a visit to Flagstaff that "few astronomers appear to realize how exceptionally excellent the seeing is in the clear dry air of Flagstaff, on a quiet night. It is so good, in fact, that a comparative novice appears to be able to see the planet more distinctly in one presentation there than Schiaparelli, at Milan, ever did."[55]

What do we make of a novice nonastronomer breezily claiming to have seen Mars more clearly than Schiaparelli, the revered (and then–still living) discoverer of the canals? His comment—produced, consumed, and circulated with a popular audience in mind—reveals the extent to which wilderness settings had come to convey scientific legitimacy. By going to the "wilderness" of the Flagstaff "mountain" observatory, a novice had learned to see and perceive the heavens on par with a great heavyweight of nineteenth-century European astronomy. It was not the author's training, skill, or intellect that had made this possible—it was the site itself.

Although this kind of representation was less explicit in the scientific publications, it nonetheless carried significant influence there, too, not least because the Mars debates began to blur the line between the popular and scientific domains. The following characterization of Lowell's authority, for instance, was written for the popular audience of *Scientific American,* but then made its way via reprint into Boston's *Commonwealth* and finally by further reprint into *Publications of the Astronomical Society of the Pacific:* "There are not more than twenty people in this Earth who have seen what [Lowell] has seen. Even some of the great observatories of the world are so situated that they have not noted the marvels which the Flagstaff observatory has revealed to us. But truth is truth, and it matters but little whether at this moment it have twenty apostles or two thousand."[56] In this author's view, the lack of confirmation for Lowell paradoxically condemned not the Arizona astronomer, but instead everyone else who was less well "situated."

In fact, as the author put it, the observatory *itself* had "revealed" the insights about Mars. The site, rather than the astronomers, carried the fundamental legitimacy for Lowell's claims about Mars.

To understand the nuances of how this idea resonated in the scientific literature, we need look no farther than the very public jostling between two mountain-wilderness observatories in the American West: Lick and Lowell. To discredit his critics at the Lick Observatory, against whom he clearly could not level charges that they were not manly enough to travel to the Western mountain wilderness, Lowell suggested instead that the men working on Mount Hamilton were not conducting themselves properly within their sublime yet challenging setting. He suggested that the Lick telescope was too large for proper observation of Mars, and that its enormous light-gathering abilities actually overwhelmed the minute definition of Martian surface features. For this reason alone, he argued, Lick astronomers could not see as many canals or perceive the importance of their changing appearances. Lowell suggested that it would be necessary for the Lick astronomers to decrease the aperture of their telescope to improve its powers of definition and take full advantage of the opportunities provided by their site. Returning to this idea again and again, Lowell faulted not the site but the appropriateness of astronomers' actions within the site. Although this characterization is somewhat different from the popular representations—which gave the site ultimate authority—it still mobilizes the same idea that the quality of astronomical work was conditioned by its geographical setting.

Campbell, at the Lick Observatory, responded in kind, leveling similar critiques against Lowell's own staff. In explaining differences between his spectroscopic results (which found no water vapor on Mars) and those performed at the Lowell Observatory (which indicated plenty of life-supporting water on Mars), for instance, Campbell argued that the Arizona astronomers probably did not understand fundamental issues related to mountain geography and that their data therefore could not be trusted. Noting that Lowell's deputy Earl Slipher had compared his twilight observations of Martian spectra to lunar spectra recorded much later in the night, Campbell commented, "Air masses at high altitudes may and usually do change rapidly. . . . It is a common occurrence for clouds to form in the afternoons in high and mountainous regions, chiefly because of convection currents which carry moisture up, for the clouds to clear away about dark."[57] In this patronizing comment, Campbell implied that Slipher did not understand mountain atmospheric conditions and had made critical scientific errors as a result. Campbell himself had done his own comparisons between Martian and lunar spectra at almost the same time to ensure that atmospheric conditions had not changed during the interval. By stressing this difference as the source of their discrepancy, he focused not on the site but on the astronomers' differing actions with their sites. Again, it was not enough just to be located in a wilderness setting; astronomers achieved legitimacy by showing that they understood and could respond to the challenges posed by the wilderness environment.

The characterizations of mountain observatories as rugged, difficult, and challenging settings thus worked hand in hand with the representation of mountains sites as sublime. The pure view available from lofty mountain peaks was augmented by the honing of intellect and instinct that astronomers experienced while responding to their challenging environments. A piercing gaze on Mars, its characteristics, and its meanings was simply not available from urban, sea-level sites like the U.S. Naval Observatory or the Royal Observatory, Greenwich. For those willing to travel to remote wilderness sites and "train" themselves in the environmental, physical, and visual procedures necessary to perceive Mars, however, the planet would reveal itself to patient scrutiny. Scientific manliness—the ability to confront and ably conquer wilderness challenges in pursuit of knowledge—was thus powerfully mobilized as a means of legitimizing various claims regarding the nature of Mars. In chapter 4 we will explore more fully the way that this form of legitimization operated in a broader context that relied on the masculine landscape gaze as a tool of both perception and power.

Science, Legitimacy, and Landscape

The episodes discussed so far in this chapter show that representations of place played a critical role both in the establishment of observatories in the American West and in the unfolding debates over Martian canals and inhabitants at the turn of the twentieth century. The legitimacy claimed and conferred by virtue of astronomers' physical presence in specific landscapes, however, also participated in several larger transitions then underway within the discipline of astronomy. Many of the foremost concerns regarding the professionalization of the discipline—the persistent participation of amateurs, the debilitating influence of popular presses and audiences, and the prevalence of human error in a science of observation—played out in the wrangling over Mars. The Mars debates highlighted and reinforced a geography of legitimacy, in fact, that directly clashed with the astronomical establishment's desire to institute universal legitimacy standards for their discipline. Leading astronomers who considered the inhabited-Mars hypothesis unimportant and unhelpful thus found themselves compelled to address the complex questions of scientific legitimacy it raised in its geographical invocations. This section examines the unsettling effect of the Mars debates on astronomy as a discipline, focusing particularly on the ways geographical representation undermined attempts to define and standardize proper scientific practice.

Scientists and writers represented the sites of Mars science in two very different modes, thus producing two very different kinds of legitimacy. On the one hand, mountain observatories were represented in terms of the purity, vision, and control available to scientists. On the other hand, mountain observatories were cast as sites of action and response, where scientists distinguished themselves by their engagement with wilderness. The distinctions between these modes of representation run parallel to the distinctions governing one of the most complex spatial divisions in the scientific world: the distinction between the field and the laboratory.[58] Laboratory science is often considered placeless, with location said to have no impact on universally replicable findings. Field science, on the other hand, is considered site-specific, in that exact results typically *cannot* be replicated from one place to another, given that local variation itself is often the topic of study. Science is practiced differently in these two spatial realms, and legitimacy is therefore cultivated differently. Traditionally, field scientists emphasize "the heroic quest of the naturalist-explorer," while bench scientists prioritize "mastery over Nature through the steady, distanced gaze of the scientist."[59]

This fundamental dichotomy between active contact and objective distance as means of accessing natural reality was both challenged and encapsulated in the complex representations that legitimized mountain observatories to produce Mars science. In the rise of observatories in the American West, the dichotomy was challenged by astronomers who gained legitimacy through representation of their practices as both controlled *and* heroic. The tropes of sublime high-altitude vision and clarity allowed astronomers to argue that they had reached a purified space above the degradations of the lower atmosphere. The new mountain observatories were thus controlled environmentally for astronomical work. At the same time, however, the tropes of scientific manliness and rugged wilderness settings cast astronomers' legitimacy in terms of the heroic interaction of specific individuals with specific landscapes. Lowell's writings, especially, relied on the idea of science as an individual pursuit. In saying that astronomers must "abandon cities and forego plains" to pursue their work on Mars, he cultivated a heroic persona of the wilderness-going scientist, whom he portrayed as more credible than any individual working in an urban setting, regardless of professional training, skills, facility, or instrumentation.

Kohler has argued that the negotiation of fundamental differences between field science and lab science—like the acceptance of amateurs and the emphasis on physical action in the field, both of which would be considered unacceptable in a lab setting—gave rise to new and vibrant sciences like ecology, which found ways to integrate elements from both sides of a spatial-scientific border.[60] In this chapter's analysis of the Mars debates, however, a different possibility emerges. An observatory is essentially a controlled space, like a lab, but its scientists pursue observational work rather than experiment like field scientists. Results and findings are theoretically *supposed* to be replicable, but the physical location of various observing sites has a significant impact on what types of observations can *actually* be made successfully. In essence, then, the astronomical observatory is a unique scientific site, in which the elements of field and lab co-mingle with no border or border zone between them. In this sense, we can say that there is no dichotomy or schism between the two kinds of science and/or the legitimacy on which they rely.

At the same time, however, this very challenge to the lab-field dichotomy, particularly in representations by Lowell, was itself challenged vigorously by leading American astronomers. The concept of an individual amateur pursuing his own agenda, developing his own standards of proof, and cultivating his own popular audiences ran against the grain of the professionalization project, which was supposed to raise the status of astronomy among the hard sciences. In using mountains and the Arizona landscape to promote his inhabited-Mars hypothesis, then, Lowell became an anathema to mainstream astronomers. Attempts to discredit him, however, were tricky. Any attempt to discredit Lowell's site would have had to acknowledge that individual sites had different characteristics. In other words, they would have been forced to acknowledge that the replicability of results—the mainstay of laboratory credibility—was impossible in astronomy. Perhaps as a result of this paradox, few attacks were made on Lowell's observatory site.

The favored method of discrediting Lowell—as a scientist and individual—however, raised the messy issue of the "personal equation."[61] If an astronomer working in a good site with good equipment could still produce bad science, as Lowell's discreditors argued, the role of human error in topic selection, observation, and interpretation was necessarily acknowledged. The "steady, distanced gaze of the scientist" was a far cry from the notion of individual eyesight, on which Lowell repeatedly tried to engage leading astronomers.[62] When Lowell offered to send eminent American astronomer Simon Newcomb medical proof of his excellent vision, in fact, Newcomb replied that he considered the subject of Lowell's eye examinations so "vulgar" that only an "utter lack of discretion" would allow it to be mentioned publicly.[63] What to do then? Critiquing Lowell as a scientist and personality was difficult for astronomers who wanted to maintain the moral high ground and avoid the popular publications in which he held strong sway. By refusing to attack his observatory, however, Lowell's critics left one of his most valuable legitimacy markers intact, for geographical location was vital to how he legitimized his theories and dismissed others' claims.

This quandary helps explain why the Mars debates ran on and on. Lowell's critics did, of course, eventually publish in the popular press in an attempt to tarnish his reputation. In the process, however, they made little progress on their desire to separate high science from popular science. In addition, the professional astronomy community's engagement with Lowell brought the issue of amateur involvement to a traumatic confrontation.[64] When the professionalization of astronomy had begun in the mid-nineteenth century, amateurs were seen as a major asset. It was widely accepted and often commented that the tasks of observation were best performed by amateurs, while theoretical work or work requiring advanced instrumentation was better done by professionals. By the end of the nineteenth century, however, the two groups were in conflict, with amateurs trying to organize their own societies, push their own agendas, and garner public interest through popular publications rather than disciplinary journals. Lowell exacerbated these tensions, claiming to be a professional but using methods common to the amateur ranks. In the debate over telescope size, for instance, Lowell championed the amateurs' argument that small telescopes were actually *better* than the professional observatories' large telescopes. As a result, the blurry distinction between amateurs and professionals had been replaced with sharp boundaries by the first decade of the twentieth century. Professionals forewent the benefits of amateur labor for the higher goal of establishing the legitimacy of astronomy among American sciences. Though amateurs continued to be involved in astronomy (as they still are to this day), they were no longer in a position to drive new developments in the discipline.

Conclusions

Throughout the period of the Mars debates, representations of mountain geography and high-altitude landscapes dramatically influenced the legitimacy of American astronomers and observatories. Individual observatory directors seemingly felt compelled to emphasize their remoteness, altitude, and proximity to mountain peaks as a means of securing credibility for their scientific work and results. In the process, they capitalized on a popular reverence for sublime mountains as sites of clarity, vision, and perspective. At the same time, they emphasized the rugged and challenging characteristics of specific wilderness sites as a way of representing the personal bravery and strength of astronomers who worked in those landscapes. In the debates over Mars, Percival Lowell was especially successful in using these tropes to promote his own credibility and to discredit his critics.

Mars astronomers' acknowledgment, celebration, and embellishment of the *site* of science raises interesting questions about the nature of the legitimacy they constructed for themselves. By emphasizing individual experience and the uniqueness of individual observing locations, mountain-based astronomers actually undermined their profession's claims to universal truth. If good astronomy could be conducted only in specific sites that served to hone the abilities of the astronomer and delimit his visual capabilities, then astronomy was essentially a field science, and replicability of results was truly unattainable. If one astronomer claimed to see something another claimed not to see (say, canals on Mars), the first could easily argue that the second astronomer's location was poorer or that his ability to conduct proper science in that location was inadequate. But there was no expectation that any other astronomer could see exactly the same thing in exactly the same way to either of the two astronomers in conflict. Neither one's results or claims could be replicated or objectively confirmed. Legitimacy on the subject of Mars became a matter almost solely of geographical representation. This fact drew enormous popular attention and raised astronomy's profile as a discipline, but not in the way the discipline's leaders had wanted.

This paradox perhaps explains some of the lingering difficulty in separating amateurs from professionals before the second decade of the twentieth century, a difficulty that allowed Percival Lowell to establish a powerful credibility for himself and his claims that Mars was inhabited. Only once American astronomy had largely abandoned its sea-level and urban sites later in the twentieth century, fully relocating to the mountains, did the geographical uniqueness of individual sites begin to lose relevance. Only then could laboratory standards such as instrumental superiority and professional standing re-emerge as primary variables in the legitimacy equation. In the era of Mars debates and the popular canal sensation, however, a metropolitan-versus-mountain dichotomy provided the critical means of differentiating among the credibility of observatories, astronomers, and hypotheses. The higher, the more remote, the more rugged, and the more sublime, the better.

FOUR

Representing Scientists: Heroism, Adventure, and the Geographical Outlook

Devoted men can always be found to undergo necessary hardships in the pursuit of scientific truth.

—American astronomer Edward S. Holden (1896)

In addition to examining the ways astronomers represented their data and their data-collection locations, we must also consider the ways astronomers represented themselves as credible individual producers of legitimate Mars knowledge. From the start of the canal debates, many astronomers conceived of and labeled their scientific study of Mars as essentially geographical, introducing an important palette of identities that colored their self-representation and their science throughout the decades of heightened Mars interest. In his widely influential 1877–78 observation report, for instance, Giovanni Schiaparelli cast himself as a geographer:

> In order to establish the topography of Mars on an exact basis, I have followed the same principles that have been adopted in terrestrial geography. A certain number of points, distinct and easy to recognize, distributed with as much uniformity as may be over the surface of the planet, creates a fundamental network for which the positions are determined with the greatest possible precision. . . . [T]he topographical description of the regions in between can be inferred without too much uncertainty from the sketches, precisely in the way that a geographer finishes the description of a country on earth by interpolating between the geometrically determined points.[1]

Also in this publication, Schiaparelli applied the term "areography" to his study of the Martian surface. This clever modification of the word "geography," which inserted the Greek name for Mars ("Ares") in place of the Greek name for Earth ("Geos"), quickly became the standard term for Mars science.[2] Despite numerous post-1877 developments in the scientific understanding of and approaches to Mars, Percival Lowell launched his own career as a Mars scientist twenty years later with similar claims of a fundamental connection between his work and geography, saying "areography is a true geography, as real as our own."[3]

The durable representation of Mars astronomy as a kind of geography allowed astronomers to traffic profitably in representations of themselves as field geographers. The use of cartographic representation (discussed in chapter 2) and the prioritization of mountain observatory sites (chapter 3) were thus part and parcel of a broad representational project that cast astronomers in a heroic light. In an era when popular audiences' attention was drawn by any mention of travel or exploration, Mars news ran in the same magazines and newspapers as reports and narratives from terrestrial expeditions, reflecting editors' acknowledgment of a perceived connection between astronomy and the field sciences.[4] In the November 19, 1910, issue of *Scientific American*, for example, W. E. Rolston's "Mars During the Recent Opposition" was followed by an unsigned article titled "Field Work in Botany."

Percival Lowell actively leveraged this perception by making direct comparisons between Mars astronomers and Earth's celebrated polar explorers. An analogy had long been asserted between the prominent white patches at the Martian poles and Earth's own polar ice caps, critically providing Lowell and other inhabited-Mars advocates with an obvious source of water for their predicted canals.[5] Lowell also cleverly used the Martian poles to anchor a rhetorical move that strongly supported his legitimacy as a scientist and interpreter of Mars. By claiming an unimpeded view of the Martian poles, he crowed that astronomers had achieved a long-sought terrestrial triumph "at much less expense and at absolutely no hazard" than the heroic polar explorers themselves. Asserting his own superiority over the many failed expeditions in terrestrial Arctic and Antarctic regions, Lowell quipped, "There are advantages in thus conducting polar expeditions astronomically. One not only lives like a civilized being through it all, but he brings back something of the knowledge he went out to acquire." Referring to the act of canal discovery as a "stirring" event, "akin to the thrill of finding unknown land in our own Antarctic regions," Lowell presented Mars science as a form of heroic exploration with which many popular audiences were extensively familiar.[6] News from the Arctic provided "a focus for patriotism, even jingoism on occasions, and the resulting national heroes, with their stories of hardship, triumph, and occasionally disaster, were especially appealing to a Protestant popular imagination."[7]

These comparisons were accepted enthusiastically by popular commentators. In the *National Review*, for instance, a writer introduced the topic of Mars to his readers in this typical way: "Astronomers are the explorers in this case, and by their telescopes they have been able to find out much more concerning the southern frozen seas of Mars, which, at its nearest, is thirty million miles away, than is known of our own Antarctic regions."[8] Writers and editors also sometimes directly linked Mars news and polar expedition reports, indicating the extent to which Lowell's comparison influenced generalist audiences. Popular science writer E. T. Brewster, for instance, opened his *Atlantic Monthly* review article on "The Earth and Heavens" with a discussion of the Peary and Scott Arctic expeditions. He then continued without transition:

> There seems to be no need for either Pearys or Scotts among Mr. Lowell's Martians. Our nearest planetary neighbors ought to know their flat and sea-less world far more completely than the children of men know theirs. In fact, even our own maps of the Martian surface have no tantalizing blank spaces at top and bottom, while, thanks to the nearly complete annual melting of its snow-caps, the poles of that other world are as familiar to the inhabitants of both as are the regions between. A mountain on Mars a quarter of the height of unknown peaks in Alaska and Antarctica or on the Roof of the World would have been seen years ago. A few miles of perpetual ice prove to be a more impassable barrier than sixty millions of empty space.[9]

The linking of Mars science with polar science thus not only contributed to a growing perception of astronomers as adventurous heroes, but it also concentrated popular interest specifically on Mars' polar landforms. Given that Lowell's inhabited-Mars hypothesis depended fundamentally on the idea that Mars was irrigated by seasonal snowmelt originating at the polar caps, this element of the popular discourse produced strong support for his position.

The association of Mars astronomy with field sciences, however, was not merely rhetorical or representational. Astronomers also borrowed from field-based sciences methodological and evidentiary standards that deeply influenced the conduct and nature of their scientific work, including its results. In addition to *representing* themselves as geographers or explorers, it seems, many astronomers were actually *thinking* about their work as geography or physical exploration. This chapter examines the reliance of individual astronomers on hybrid conceptions and representations of themselves as astronomer-geographers who were forced by their subject matter to pursue new approaches to their science. In particular, it examines a suite of scientific expeditions in which American astronomers transported their instruments and personnel to various high-altitude locations in order to gather data about Mars. Publicity surrounding these expeditions and their findings employed many of the strategic representations of mountain sites that were discussed in chapter 3. In this chapter, however, I want to focus on the ways that expeditionary travel allowed astronomers to represent themselves as masculine heroes, adventurous travelers, and scientific explorers. Given that several critical debates about Mars were alternately enflamed or quieted by data gathered outside the walls of formal observatories, the representational and methodological maneuvers that emerged from expeditionary movement played an important role in the production of Mars knowledge.

Gazing on the Martian Landscape

Astronomers established a fundamental connection to the science of geography both through their methodological decisions and in their textual descriptions of these methods. In some cases, including the two quotes from Schiaparelli and Lowell above, this connection was direct and explicit. In many cases, however, astronomers who never made any explicit reference to geography were nonetheless perceived by popular audiences to be participating in a fundamentally geographical investigation of Mars. The reasons for this persistent linking become clear if we look at the two methodological elements most often invoked in the study of Mars: (1) direct optical apprehension of the planet's surface as the primary means of collecting data, and (2) intuitive visual analogy as the primary means of interpreting the Martian landscape. As Gillian Rose articulated in a now-classic essay on the history and practice of geography, these are the same methods that have long defined the essence of geography as a scientific and practical discipline.[10] Although visual methods also shaped other scientific disciplines during the same time period, this section focuses closely on astronomers' use of specifically geographical methods because so many Mars observers (and their audiences) compared Mars science to geographical science.

Particularly during its formative years in the late nineteenth century, geography came to depend for both its data and its analyses on a visual encounter between the subject-geographer and the object-landscape. Geographers went into a landscape to look at it, to see with their own eyes its dimensions and nuances, and to visually assess the relationships and processes that had given rise to its unique characteristics. For the geographer in the service of the government, the academy, or private enterprise, knowledge has long been constructed primarily through individual acts of observation. The geographical way of knowing is therefore inherently also a way of seeing.[11] Even those "invisible" forces that impact the relations between humans and environment (e.g., religious belief) can be seen in their influence on the observable landscape. The nature of geographical investigation, therefore, requires the geographer first and foremost to gain a direct view of the area under study, typically by going there and establishing a physical vantage point. The optical processes involved in "looking" at a landscape, however, are not alone sufficient for understanding it. What is seen with the eyes is merely raw data. It is in the act of "gazing"—when "the eye [holds] the landscape together as a unit and the geographer then analyse[s] the view, selecting features requiring elucidation"—that geographic knowledge is produced.[12] Geographic practice thus relies both on the movement-into-viewing-position of a scientific observer and on a conflation of that observer's visual and mental operations to produce an understanding of a landscape.

Despite the difficulties of actually "seeing" the red planet from 35 million miles away at its closest, analysis of astronomers' publications shows that many of them used (or claimed to use) a methodology best described by Rose's term: "the geographer's gaze." The work of Percival Lowell shows perhaps most clearly how the use and articulation of this geographical method came to shape the debates over Mars and the self-representation of those astronomers who engaged in them. Despite Lowell's interest in telescopes, his discussions of the Martian landscape often implicitly removed this fundamental instrument of vision from the reader's consideration. In his influential first book, *Mars*, for instance, Lowell claimed astronomers could "gaze upon the actual surface features of the Martian globe," implying that it was the naked eye rather than chromatic screens or reduced apertures that brought these features into view. In a popular article on Mars' polar caps, likewise, he asserted a direct view in reports that he had found a polar sea: "It lies in a valley between two mountain ranges. Of this we are almost as sure as if we had climbed one of the enclosing summits and looked down upon it."

Lowell was not the only one to marshal this perspective in his writing, even if he was the most explicit. Writing in the *North American Review*, for instance, popular French astronomer Flammarion claimed, "with our own eyes we see the polar snows melt during the summer and reappear in the winter." This narrative voice, which asserted a direct gaze upon the Martian landscape, was extremely powerful in its ability to persuade popular audiences that Mars was an inhabited planet. Lowell's opponents reluctantly acknowledged its contribution to his rising popularity, grumbling that his "readable style . . . gives an impression that he has been there and seen it all." No matter how frequently they cautioned popular audiences and other scientists not to jump to conclusions about Mars, however, these critics found it difficult to combat the idea that legitimate knowledge was produced through a direct gaze upon the Martian landscape.[13]

During the escalation of the Mars canal debates, claims of a direct view of Mars maintained legitimizing power only if accompanied by a complementary and credible interpretation of the Martian landscape's deeper meanings. It was not enough to "see" lines on the surface of Mars without also offering an explanation of the landscape processes that were likely to have produced those lines. In the competition to build legitimacy for their various explanations of the Martian landscape, astronomers who represented their interpretive methodology as analysis-while-looking found the most success. Lowell again provides the most detailed example of the use and representation of this visual interpretation methodology to mobilize legitimacy for claims about Mars. Although he was not the only astronomer to use a visual interpretation method, he wrote explicitly and at length about his conception of "the mind's eye" as an interpretive tool. Essentially, Lowell's "mind's eye" was an abstract quality of perceptiveness or intuitive recognition that relied on an understanding of Earth's landforms and processes to develop terrestrial analogies that he then applied to Mars as definitive. Commenting on observed changes in the appearance of Mars, for instance, Lowell argued that they led him to conclude "that Mars is a living world, subject to an annual cycle of surface growth, activity, and decay; and shows, in the second place, that this Martian yearly round of life must differ in certain interesting particulars from that which forms our terrestrial experience."[14]

Using the "mind's eye" as methodology, Lowell cleverly dismissed his well-funded detractors by arguing that the ability to perceive and make sense of landscape detail—a personal skill that was not possessed by every astronomer—was more important than telescope size. Using a mid-sized telescope for his own work, he thus turned the attack against those who were using the world's largest telescopes, accusing them of failures in interpretation. According to Lowell, some astronomers (himself included) could mentally comprehend the nuances of a landscape without much technical support, while others had to be told what to look for or shown how to see what was right before their very eyes. "Most people see only what they are prepared to see; as is well instanced in astronomy by those observers who manage to mark with surprisingly small instruments what others have already discovered, and yet who make no discoveries of their own."[15] The "mind's eye" thus functioned for astronomers and their audiences as a geographer's gaze that allowed astronomers to intuitively piece together the visible elements before them, simply by virtue of looking with an open mind.

Lowell invited his audiences to engage in the gaze as well, further cementing his argument that it was not instrumentation or technical training that produced the most legitimate understanding of Mars. Presenting his groundbreaking (though extremely small) 1907 photographs of Mars, for example, Lowell conjured a fascinating world open to the geographer's gaze (refer to fig. 2.14 above):

> One thing he who scans these circles must understand, or he will miss the full measure of the wonder they contain. His brain must be open to them; not his eye alone. For what is before him is no meaningless articulation of black and white, but the portrait in its entity of another world, imprinted there by that world itself. Sharp set against the black of space this circlet of light displays to him an earth, comparable in grandeur and self-containment with that on which he dwells. Small to the sight, in the brain it takes on its true dimensions, and to the mind's eye becomes the globe it really is, which, could he find himself transported thither, would seem the essential sum and center of the universe, as now to most men our own world comprises all they know.[16]

Some of Lowell's allies in other disciplines echoed the point, arguing that an open mind was more important than astronomical training for the study of Mars. Edward Morse, Lowell's zoologist friend and director of the Peabody Academy of Science, wrote:

> A student familiar with a general knowledge of the heavens, a fair acquaintance with the surface features of the Earth, with an appreciation of the doctrine of probabilities, and capable of estimating the value of evidence, is quite as well equipped to examine and discuss the nature of the markings of Mars as the astronomer. If, furthermore, he is gifted with imagination and is free from all prejudice in the matter, he may have a slight advantage.[17]

Lowell's discussion of observer perception or intuition functioned largely to legitimize the work of those who claimed to see canals on Mars and to delegitimize the work of those who claimed to see no geometric landforms at all. At its core, however, this rhetorical maneuvering relied heavily on the underlying condition for and the ultimate goal of the geographer's gaze: intuitive understanding of a directly viewed landscape.

Method, Movement, and Identity

The use of a geographer's gaze was a radical departure from typical practice in the discipline of astronomy. Before the Mars-canal era, celestial inquiries were mainly approached with theoretical methodologies drawn from mathematics and physics. With regard to the determination of Mars' temperature, for instance, computational analysis of the planet's mass and orbital trajectories predicted that Mars should be considerably colder than Earth—probably never above freezing—given that it was smaller and further from the sun. (Today's science tells us this is indeed the case.) Introduction of visual interpretation, however, changed the conventional wisdom on this issue. Visual observations of the north and south poles of Mars had long revealed large white patches that appeared to enlarge in Martian winter and shrink in Martian summer (see fig. 4.1). Suggestive of the appearance of polar snow and ice on Earth, this visual evidence from the red planet intuitively suggested a seasonal melting of ice that would confirm Mars' average temperatures to be considerably above freezing, at least during the summer.[18] Despite some protests that unproven hypotheses about the white patches should not be allowed to negate sound theoretical predictions about extreme cold on Mars, the "melting" of the "polar snows" was widely accepted as conclusive observational evidence that Mars had a temperature comparable to Earth's.[19] During the height of the discussion over Mars' hospitability to life forms, the belief in a relatively warm Martian atmosphere contributed strongly to the arguments of those who favored the view that Mars could support life. Even neutral Schiaparelli offered that "as far as we may be permitted to argue from the observed facts, the climate of Mars must resemble that of a clear day upon a high mountain."[20] The geographer's gaze thus allowed, and even required, astronomers to piece together their understanding via visual interpretation rather than calculation or numerical analysis.

Another debate, closely related to the temperature question, likewise hinged on the methodology of the geographer's gaze. Given that the polar caps were believed to melt (based on visual interpretation), many astronomers logically assumed the planet must have liquid water on its surface at various times throughout the year. When spectroscopic tests failed to show any polarized light reflecting from dark areas on the Martian surface, American astronomer William Pickering proposed that they could be vegetation

PLATE XXIX.

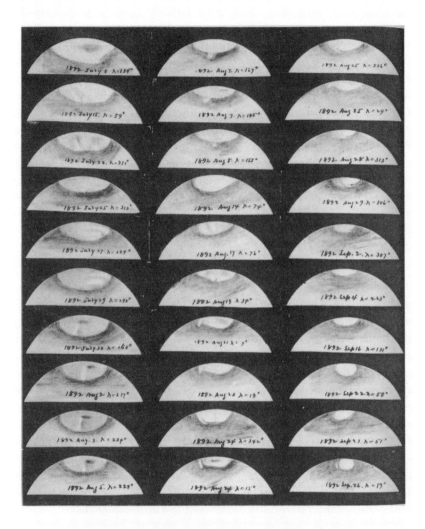

THE SOUTH POLAR CAP OF MARS IN 1892 WITH THE 12-INCH AND

THE 36-INCH REFRACTORS BY E. E. BARNARD.

Figure 4.1. Published sketches of Martian polar cap variation, 1892. These sketches, compiled and published by American astronomer E. E. Barnard, were intended to chronicle a seasonal change in the Martian surface appearance that had long been observed by numerous astronomers: the waxing and waning of a white spot in the planet's southern polar region. This plate appeared in the June 1895 number of *Popular Astronomy* (facing p. 440).

instead of oceans. This theory, which shortly became central to Lowell's inhabited-Mars hypothesis, rested on visual observations of the patchiness and variability in the colors of Mars' surface.[21] Although subsequent spectroscopic analyses were inconclusive in determining whether even the water vapor necessary for vegetative growth existed in Mars' atmosphere, the new vegetation theory achieved widespread acceptance because it made visually intuitive sense as an explanation for the mottled "green" areas on Mars.[22] Even the reddish areas could be explained as vegetation: "there is certainly no impossibility in the conception that vast forests of some such trees as copper-beeches might impart to continental masses hues not unlike those which come from Mars."[23] Again, this rhetoric and logic prioritized landscape-level observational analysis over theoretical or experimental findings. Once subjected to evidentiary and methodological standards adopted from the field sciences, the planet Mars could no longer be described decisively with theoretical or mathematical predictions about its conditions.

As Mars scientists strategically adopted these new methods from field-based disciplines, they also paradoxically began to associate Mars science very closely with the idea of travel. As Rose and others have noted of the geographer's gaze, "the practice of fieldwork was crucial to this scientific understanding of landscape."[24] In order to understand a landscape, the scientist needed to go into the field, both to gain a direct view and to perform the visual interpretation required for true comprehension. The gaze thus required some bodily movement of its practitioners. For geographers, the nature of fieldwork followed an obvious formula, despite its difficulty and expense: geographers raised funds, selected a field crew, packed equipment and supplies, and then sailed, rode, hiked, caravanned, paddled, were carried or otherwise made their way into the landscapes they wanted to study. For the astronomers studying Mars, however, the prospect of fieldwork was a bit more complicated. Space travel was little more than a fantasy at the time (although some astronomers did in fact engage in vivid imaginations of an actual journey to Mars through fictional pieces they considered separate from their scientific work),[25] and few astronomers went as far as Percival Lowell in their mainstream scientific writing to evoke a sensory travel experience for readers:

We may thus make a far journey without leaving home, and from the depths of our arm-chairs travel in spirit to lands we have no hope of ever reaching in body. We may add to this the natural delight of the explorer, for we shall be gazing upon details of Martian geography never till last summer seen by man. . . . We will begin our journey at the origin of Martian longitudes and travel west, taking the points of the compass as they would appear were we standing upon the planet.[26]

More commonly, astronomers augmented their claims of direct landscape perception by pursuing *terrestrial* travel as a proxy for actual fieldwork on Mars. In search of Mars data, prominent American and European astronomers regularly engaged in expeditions to mountains, deserts, and islands, often in the tropics and the Southern Hemisphere. Lugging their equipment and personnel with them in movable field observatories and boasting improved views of the heavens, these men successfully cultivated reputations as travelers and explorers who went directly into the field to gather data.

The remainder of this chapter will examine the ways various American astronomical expeditions mobilized the gaze-as-methodology and the gaze-as-representation to sensational effect. Publicity surrounding astronomical expeditions to southern mountains caused huge swings in popular opinion regarding the geography of Mars. As these expeditions, their leaders, and their findings became focal points in the turn-of-the-century legitimacy wars over Mars, methodological and representational maneuvers served to legitimize the expeditions themselves and also to cultivate important identities for individual knowledge producers. Although the subject of astronomers' geographical writing (Mars) and the actual location of their travels (Earth's lower-latitude mountains) did not coincide, gaze-related methods of inquiry and representation powerfully fused these disparate locations in remarkably coherent and convincing narratives. In this regard, turn-of-the-century Mars astronomy can and should be examined closely as a kind of field geography.

The New Heroes: William Pickering in Peru

One of the early associations of Mars science with terrestrial travel developed somewhat surreptitiously. Astronomer William Pickering, who was sent to Peru in 1891 to photographically map the southern skies for Harvard College Observatory, spent most of his time observing the planet Mars instead of carrying out his assigned duties. Actively defying repeated orders given to him by the observatory's director (who happened to be his older brother, Edward), Pickering insisted that the opportunity to observe Mars from such a sublime high-altitude location should not be missed. During the perihelic opposition of 1892, he observed Mars extensively and published findings—mainly in daily newspapers and popular magazines—that challenged the conventional wisdom on Mars. Pickering validated this Mars work with images of dramatic mountains and accounts of heroic mountain-climbing that established a close representational association among Mars science, masculine heroism, and the geographer's gaze. In the process, he spurred the first widespread American popular interest in Mars and initiated an era of Mars-related scientific travel to tropical, high-altitude sites.

Ironically, Harvard's southern station was never meant to study Mars. In the 1880s Harvard's observatory director, Edward Pickering, had aggressively sought funding to establish several high-altitude sites outside of Cambridge.[27] Edward's was one of the loudest voices advocating for the establishment of observatories at high altitude, and he did not intend for his own urban, sea-level observatory to be swept aside in the tide of enthusiasm for mountain observatories (see chapter 3). Upon being granted control of a very large bequest that came to be known as the Boyden Fund, Edward immediately worked to implement his visionary plan to establish two high-altitude sites under the auspices of Harvard. One was to be located in the American West (although this plan never came to fruition), and another would be located in South America.[28] The Southern Hemisphere observatory would have the unique advantage of being able to see the southern skies, which were invisible from the Cambridge facility. As Edward saw it, this new facility needed only limited instrumentation and a small staff. The observers would focus all their time and effort on collecting photographic and spectrographic data, with the resulting glass plates sent in crates back to Harvard. Computation of the data and analysis of results would then be done by Harvard's existing staff "computers" under Edward's direction in Cambridge.[29] The new southern station was thus envisioned as little more than an outpost.

To establish the station as thus envisioned, Edward sent Harvard staff astronomer Solon Bailey and a small expedition party to investigate possible sites in the Andes Mountains and along the western coast of South America between 1889 and 1891. Bailey's descriptions of his journeys, activities, and impressions of Central and South America were presented in most detail in a travelogue-style narrative that served as an introduction to his formal discussion of the expedition's scientific work. His descriptions were also reported in various mainstream scientific and popular publications.[30] These publications established several important representational precedents for Pickering's work, capitalizing on the expectation that Harvard's "Boyden Station," as it eventually came to be called, would function as a frontier outpost rather than a formal facility. Bailey characterized the station in terms of its isolation from civilization in rugged environments that required tremendous fortitude and effort from the field-going astronomers. The heroism of these astronomers and their commitment to difficult fieldwork thus became a dominant theme in the way the observatory was represented, well before it undertook any work on the planet Mars.

First establishing a temporary facility in 1889 on an unnamed Peruvian mountain that they dubbed "Mount Harvard," Bailey's small expedition reported finding their situation extremely challenging, as they housed themselves in a shack with paper walls and were "almost entirely cut off from communication with the world below, and all supplies, even including water, had to be carried by mules from the valley below, a distance of about eight miles." Despite these significant discomforts, the group endured its lot for the first observing season, presenting an image of hardy astronomers taking the "ideal hardship" and challenges of fieldwork bravely in stride.[31] During the cloudy season, Bailey and his assistant (his brother, Marshall Bailey) continued the search for a permanent site, exploring several locations in Peru and Chile. They eventually settled on a small village near the city of Arequipa, Peru, which offered greatly improved convenience, including comfortable housing in the village.

Despite the improvements in the new site, Bailey and his successors continued to represent the outpost as a rugged site of heroic fieldwork. In these representations, the focus turned from the observatory site itself to the surrounding mountains. In describing his first investigation of the Arequipa site, Bailey made the following report:

> The first view of the city is really beautiful, surpassing in picturesqueness any other Peruvian city we had seen. . . . Above the city, which rests just at its foot, rises the "Volcano of Arequipa" El Misti, a nearly extinct volcano about nineteen thousand feet high. This symmetrical cone-shaped mountain is flanked toward the east by Pichu-Pichu and to the west by Chachani, the latter rising to the height of twenty thousand feet, and usually covered with snow.[32]

The mention of surrounding mountains and their peak heights quickly became a standard trope in the representation of the Arequipa observatory. Photographs of the observatory station always included a dramatic peak in the background, identifying the new facility with dramatic landscapes and majestic mountains. Often, photographs from the observatory site actually omitted the buildings in favor of the mountains (see fig. 4.2). Although the observatory was actually on a plateau rather than a mountain, it was usually considered a mountain site on account of these representations, (just

FIGURE 12.—MT. CHACHANI, FROM THE AREQUIPA OBSERVATORY.

Figure 4.2. Photograph taken from Harvard's Boyden Station, 1892. This was one of many plates that appeared in the August 1896 number of *Publications of the Astronomical Society of the Pacific* (following p. 245) to illustrate European and American observatories. Although it shows the landscape around Harvard's remote Peruvian observatory dwarfed by the Andean Mount Chachani, it does not actually show the observatory itself.

as later also occurred for the Lowell Observatory; see chapter 3). In Holden's monograph on the conditions at European and American "mountain observatories," in fact, the Boyden Station received considerable attention. It was the first station mentioned in the chapter on "The Observatories of South America," with the second sentence of the entry declaring: "Fourteen miles from Arequipa is the mountain Chachani (20,000 feet) which is always snow-capped."[33] The surrounding mountain landscape thus became critical to representations of the Boyden Station, marking it as a dramatic site that influenced the nature and quality of the science conducted there.

After Bailey's exploratory mission, Pickering was sent to Peru to establish the photographic program of the Arequipa outpost. Almost immediately upon arriving, however, he turned his attention to nonscientific exploits such as climbing the surrounding mountains. As an avid mountaineer and founding member (like his brother Edward) of the Appalachian Mountain Club, Pickering relished the prospect of tackling such "enticing summits so near at hand."[34] He immediately targeted El Misti and determined to at-tain its summit, even before Bailey had left for his return trip to the United States. Although the 1891 ascent of El Misti had absolutely no scientific goals and was little more than a recreational endeavor (Pickering apparently carried neither a camera nor any meteorological equipment on the trip, despite using Harvard funds and taking Harvard staff members as his team), it clearly enhanced the scientific reputation of the observatory.[35] Holden, for example, gave mountain-climbing expeditions prime billing in his description of the Boyden Station. Drawing on Pickering's published account of ascents of both El Misti and Chachani in *Appalachia* (the journal of the Appalachian Mountain Club), Holden wrote breathlessly of the grave attacks of mountain-sickness that had left various members of the El Misti climbing party "prostrated" or "delirious," rendering everyone but Pickering and several native guides unable to reach the summit.

> We found that all persons with blood of the white races in their veins were subject to the complaint, the pure-blooded Indians only being more or less exempt. Half-breeds who had spent all their lives in Arequipa were often more susceptible to it than ourselves. In my own case this susceptibility rapidly wore off and after my first night on the Misti I never again felt any very serious inconvenience.[36]

Pickering was so successful at establishing a heroic association of mountain-climbing and scientific legitimacy at the Boyden Station that it persisted even after his tenure ended. After a period of long conflict with his brother Edward (described below), Pickering was removed as the Boyden Station's director, and Bailey returned to Peru to replace him. Having been one of those members of the party that failed to reach El Misti's summit in 1891 when Pickering alone succeeded, Bailey determined to try again. Setting as one of his first orders of business the establishment of a functioning meteorological station on El Misti's peak, Bailey was perhaps most interested in proving himself, both to the man he was replacing and to the man who had entrusted him with responsibility for the observatory. Bailey wrote dramatically of his desire to conquer the volcano, revealing a powerful entanglement of scientific interest and the romantic pursuit of heroism: "El Misti stands alone. At first a sort of awe kept me from considering as possible the establishment of a station on its summit; but always, as I looked upon it, the impulse became stronger and stronger, and finally it could not be resisted." After numerous exploratory missions and the construction of a mule trail on El Misti's slopes, Bailey succeeded in establishing "the highest meteorological station in the world," which was lauded in the popular press not only for its potential contributions to science but also for the heroics involved in its installation.[37]

These attempts on El Misti became central to the identification of Harvard's southern outpost before it became known for producing knowledge about Mars. The fact that astronomers had traveled to a remote tropical location and undertaken difficult ascents of high mountains became fused with the fact that they were also carrying out astronomical observations. Although their mountain-climbing was actually unrelated to their astronomical mandate, Pickering and Bailey gained scientific legitimacy through the persona of the heroic mountain-climber. This occurred despite an admitted conflict between the physical strains of high-altitude climbing and an ability to carry out reliable scientific work upon high Andean summits. As Bailey wrote to Edward of his need to rely on mules to achieve El Misti's peak, "A little of the romance of mountain climbing is perhaps lost by using mules legs, in part, instead of depending entirely upon ones own: but what is lost to sentiment is many times made up to science, for one arrives at the summit with quiet nerves and in good condition for exact work."[38] Bailey thus removed any embarrassment he may have felt regarding his physical abilities as a mountain-climber, offering scientific achievement (the installation of meteorological instruments) as an equal form of heroism. Likewise, though some considered impossible the prospect that astronomers would ever be able "to live and work at these great altitudes," Holden used Pickering's fortitude as a heroic example to the contrary, stating: "Devoted men can always be found to undergo necessary hardships in the pursuit of scientific truth."[39] The persona of the heroic masculine mountain-climber was thus easily merged with the persona of the gazing field scientist in ways that enhanced the legitimacy of the scientific claims emanating from this early astronomical outpost.

With all the emphasis on scientific heroism and dramatic conquests of fabled Andean mountains, the observatory's original mandate to undertake a program of photographic mapping of the southern skies was barely mentioned in the popular press. During the two-year period that Pickering directed the station, the staff actually carried out very little photographic work. Owing to Pickering's strong interest in planetary astronomy and his devotion to visual (rather than photographic) techniques, he essentially ignored instructions from Cambridge and pursued his own program of Mars observations.[40] Although this choice led to bitter dispute between the Pickering brothers and eventually ended in Pickering's dismissal, it ensured that the heroic mountain conquests that came to define the Arequipa outpost also became closely associated with representations of Mars science.

In 1892 the red planet made a remarkably close approach to Earth during a perihelic opposition. It was the first such opposition since 1877, when Schiaparelli had discovered Martian canals and Asaph Hall had discovered Mars' two moons. Pickering would not waste what he considered a glorious opportunity to contribute to knowledge about Mars, given the "splendid atmosphere" above the Andes at his "remote and isolated position."[41] Pickering claimed that it was these perfect atmospheric conditions themselves that enticed him to study Mars and its enigmatic markings, despite Edward's repeated requests that he begin the photographic work for which the observatory had been equipped. Pickering's major contribution to Mars studies as a result of his 1892 and 1894 observations (the former conducted in Arequipa) was a revision of the assumption that the dark areas on the Martian surface were bodies of water. Although others addressed this question through spectrographic analysis (discussed later in this chapter), Pickering challenged the dominant theory purely on the basis of visual observation. Noting that the dark areas on Mars were mottled and patchy, with some linear features apparently crossing through them, Pickering argued they were less likely to be oceans than tracts of vegetation. He also discovered dozens of circular dark spots that he initially labeled as "lakes" but later modified to be "oases" in accordance with his vegetation theory. Pickering's scientific claims thus gave considerable traction to Schiaparelli's original canali by offering more plausible explanations for both their extraordinary width (the dark lines were reconceived as bands of vegetation growing alongside linear waterways rather than the waterways themselves) and their links to nonlinear features (the dark lines crossing through mottled areas were said to be canals passing through low-lying swamps or forests). Despite challenges to Pickering's findings, the dark areas on Mars were never again seriously thought to be oceans. Even those who used spectroscopy in an attempt to disprove the existence of water switched to focusing on the likelihood of water vapor in the air rather than surface water in lakes or ocean basins. Pickering achieved this remarkable influence on Mars science largely through implementing a geographer's gaze in both method and representation. Methodologically, he relied on the direct visual acquisition of data from a sublime vantage point. Representationally, he used his climbing exploits to cement the ultra-legitimate persona of the field scientist in pursuit of a clear and direct view.

Pickering's rising influence on the question of Mars, however, grew in stark contrast with his diminishing status at the Harvard College Observatory, within his own family, and among the ranks of professional astronomers. In willfully disobeying his scientific orders, Pickering provoked repeated admonishments from Edward about the misuse of observatory equipment, time, and funds. Correspondence shows that Pickering chafed at his inferior position in the observatory hierarchy, considering himself to be a full-fledged "director" of a permanent observatory rather than a temporary caretaker of an expeditionary outpost, as Edward had evidently intended.[42] In an attempt to prove that his efforts were worthwhile, Pickering often reported his Mars-related findings directly to the popular press, sending regular cables to the sensational *New York Herald*. Although his claims about lakes and vegetation on Mars mystified a number of major American astronomers (thus further aggravating his relationship with Edward, who had begun to receive skeptical or critical correspondence from associates), Pickering became even more heroic in the popular press.

In large part, it was the distance from Cambridge—both physical and perceptual—that allowed Pickering to assert his independence, which surely would have been impossible if he had stayed at home. As recent scholarship in historical and cultural geography has shown, the act of physical movement via travel can destabilize subject identities in complex ways.[43] In Pickering's case, expeditionary travel allowed him to challenge his inferior identity as Edward's younger brother. From languishing in his brother's shadow at home, both as an astronomer and as a mountaineer, Pickering was transformed into a powerful hero in the Andes. Edward may have been president of the Appalachian Mountain Club, but Pickering summitted El Misti—a peak that dwarfed the highest of those in the tame Appalachian belt—in a grueling ascent that saw all the nonindigenous members of his party succumb along the slopes. Edward may have been director of a powerful astronomical institution, but Pickering had turned his telescope toward Mars from a new vantage point so sublime that it practically insisted upon the study of the poorly known planet. Pickering's heroic travel to and fieldwork in a new region with new geographical and climatic conditions thus allowed him some freedom from his brother's shadow. Edward's critical letters served as reminders of his old status at home in Boston, but his glorification in the *New York Herald* revealed a new subject identity that was possible only because he had changed location.

Although Edward eventually recalled his brother to the United States and replaced him with Solon Bailey, the Boyden Station's association with Mars and Pickering's association with Mars and mountain heroics would persist. In just two years, Pickering had set the stage and thrown down a gauntlet for future Mars observers. Southern mountains came to be seen as ideal sites for Mars investigation, owing to their improved atmosphere and unique perspective on the southern sky. Furthermore, the heroism associated with remote fieldwork came to define the persona of the successful Mars scientist, as we will see. Not only did mountain-related expeditions give astronomers the opportunity to claim a heroic gaze on Mars, but the associated distance from civilization allowed them to destabilize and challenge accepted norms of scientific conduct. Sending cables directly to the popular press made more sense for traveling astronomical expeditions, where the only other alternative was to delay the announcement of scientific discoveries while correspondence made its weeks-long steamer journey back to the discipline's power centers. Travel and distance—the purview of the gazing field scientist—thus changed the dynamics of how Mars news was created, circulated, and consumed. The two-year period during which Pickering focused the Boyden Station's reputation and resources on Mars science may have been brief, but it had a significant impact on the way the most successful Mars astronomers represented themselves and cultivated their legitimacy from that point forward.

The New Travelers: David Todd in Chile

After his recall from Peru, William Pickering was given leave from Harvard to accompany Percival Lowell's expedition to establish an observatory in Arizona. Although Pickering stayed in Flagstaff for only the 1894 observing season, the Lowell Observatory soon mastered Pickering's techniques of representing astronomers as heroes, bypassing establishment publications, and manipulating the distance induced by travel to destabilize traditional hierarchies within the astronomical establishment. These techniques produced complex and sometimes competing forms of legitimacy. Representing oneself as a scientific hero enhanced professional standing to some extent but also created the danger of a paradoxical loss of legitimacy as other scientists began to perceive the overindulgent hero as little more than a popular or sensationalist figure. Lowell crossed this line even more quickly than Pickering, never achieving widespread acceptance within the community of professional American astronomers as a result. In return, however, Lowell enjoyed national and international prominence on the basis of his representational tactics.

In 1907, after more than a decade of arguing for the existence of Martian canals, Lowell conceived a South American expedition that he hoped would allay astronomers' mounting criticisms of his work. Appointing veteran solar-eclipse expeditioner David Todd of Amherst as director of the expedition, Lowell sent a small party from Flagstaff to the Andes Mountains to observe and photograph the surface of Mars during a favorable opposition. This was the closest approach of Earth and Mars since 1892, when Pickering had made his Arequipa observations. Lowell's stated intent for the expedition was to capture definitive photographic evidence of the Martian canals, thus proving incorrect his critics' favored explanation of Mars' linear surface appearance as the product of optical illusion. He hoped that the new images would remove issues of instrumentation, location, and atmospheric quality from the equation, noting that "a photograph can be scanned by everybody, and the observation repeated until one is convinced."[44]

With funding and scientific instructions from Lowell, Todd packed up Amherst's new eighteen-inch refracting telescope in May 1907 for a journey to the western coast of South America.[45] Accompanied by a technician from the firm that had built the telescope (Albert Ilse), an assistant from Amherst (R. D. Eaglesfield), an astronomer from the Lowell Observatory (Earl Slipher), and his own wife (Mabel Loomis Todd), Todd traveled by steamer from New York via the isthmus of Panama. He selected his South American point of disembarkation on the basis of weather reports he gathered at various ports of call, eventually setting up the telescope at Alianza, Chile, on the edge of a mining company's isolated workers' settlement (see fig. 4.3). From June 23 to August 1, the party made telescopic observations on a nightly basis under such clear atmospheric conditions that Todd labeled Alianza "an astronomer's paradise." Although the expedition made some observations of Saturn and of an annular solar eclipse, Todd reported as its "crowning success" Slipher's capture of more than 9,000 photographs of Mars. Lowell would later use these photographs as evidence of intelligent life on Mars, with Todd heartily supporting the theory after seeing from Alianza canals he described as "positively startling in their certainty."[46] Todd and company packed up their instruments and photographic plates in August, making their way back to New York via steamer.

Although short-term astronomical expeditions were fairly common in the nineteenth century, this one was unusual. Most expeditions were aimed at seeing a celestial object or event that would be invisible from the home location. A solar eclipse that would be visible from only certain areas of the globe, for example, might require an expedition to northern Africa, or east Asia, or India simply to be in the right place to make an observation. Todd was a veteran of this type of work, having traveled around the world in search of solar eclipses. In the case of the Lowell expedition, however,

. THE AMHERST TELESCOPE IN POSITION AT ALIANZA, CHILE

The telescope was mounted in a cemented tennis-court, 4200 feet above sea-level. The large weight attached by means of a rope was for the purpose of counterbalancing the increased weight of the tube in the Southern hemisphere, it having been constructed for use at 42° North latitude. The planetary camera (not shown here) about five feet in length, was attached to the lower end of the telescope. The observing chair is in the background at the right. The members of the expedition, from left to right, are Professor David P. Todd (in charge), Mrs. Todd, Robert D. Eaglesfield, A. G. Ilse, and E. C. Slipher.

Figure 4.3. David Todd's expedition party and telescope in Chile, 1907. This photograph, which appeared with Lowell's expedition report in *Century Magazine* ("New Photographs of Mars," on p. 305), shows the eighteen-inch Amherst telescope mounted in the open air under a clear sky.

Mars was *not* invisible from North America, so Lowell clearly had other objectives in sending a party of observers over such a long distance at such great expense. Having based a decade's worth of legitimacy claims on the superior atmosphere at his site in Flagstaff, he couldn't very well argue that his expedition was seeking a clearer sky than he already enjoyed in Arizona. At best, he could make the case that Mars would be higher in the sky when observed from southern latitudes, alleviating the difficulty of "detecting its finer and more important features [from observatories in] Europe or the northern half of the United States."[47] The geometry of sightlines and the height of Mars in the sky, however, were probably much less significant than the popular sensation Lowell hoped to provoke with a high-profile expedition. In a time when scientific expeditions were especially interesting to popular audiences (and the magazine and newspaper publishers who catered to their interests), Lowell staked his reputation on the assumption that a well-publicized expedition to carry out science in dramatic mountain landscapes would be perceived automatically as a legitimate endeavor.

For short-term astronomical expeditions such as those sent to observe solar eclipses, the transport of massive equipment and numerous personnel to a remote site for even a few weeks was a major operation requiring significant advance planning. Savvy astronomers often leveraged this effort into broad popular legitimacy, depicting their eclipse-observing expeditions as grand adventures.[48] Although expedition leaders invariably published scientific reports for the astronomical journals, they also often published narrative expedition chronicles for the popular press. For any given expedition, several such chronicles might be published, including a stream of short telegrams reporting on the expedition's progress while it was underway as well as longer narrative descriptions submitted soon after the party returned home. In this regard, the Lowell Observatory's 1907 expedition to South America was just like any other. Lowell cabled the press with news from the expedition immediately upon receiving any report from Todd, and he enjoyed the development of a bidding war between several magazines seeking first publication rights to the expedition's findings. Having telegraphed a brief message of "Bravo" to Todd (in Chile) upon hearing the first reports of successful canal photographs, Lowell also sent a letter boasting of the sensation the expedition had created in the popular press:

> Bravo! I telegraphed you and bravo! I repeat. Your despatches cause our hair to stand on end and our voices to stick in our throats in true classic style. . . . The world, to judge from the English and American papers, is on the *qui vive* about the expedition as well as about Mars. They send me cables at their own

extravagant expense and mention vague but huge (or they won't get 'em) sums for exclusive magazine publication of the photographs.[49]

Taking advantage of the frenzy of excitement surrounding the expedition, both Lowell and Todd published narrative reports of the expedition and its findings, as did both Slipher and Mabel Loomis Todd.[50]

These postexpedition narratives of the 1907 Lowell expedition were so valuable to the popular press and so important to the reputations of the individuals concerned, in fact, that their publication became a matter of bitter contention between two major American magazines. Lowell and Todd had each promised his story separately to a different publisher, raising the question of whose promise included first-publication rights to the photographs taken during the expedition. For Lowell the photographs of Mars were critical to his bid to establish authority and legitimacy for the inhabited-Mars theory. As such, he did not intend to allow anyone but himself to tell the story of how they were made and what they showed. When he received a letter from *Cosmopolitan Magazine* in September 1907 requesting photographs to accompany Todd's expedition narrative, Lowell replied angrily that Todd had no authority over the photographs. In fact, he had already negotiated an exclusive arrangement with *Century* magazine for an article "giving first publication by anybody of the South American results," including the photographs, narrative, and interpretation. For Todd as the leader of the expedition and for his editors at *Cosmopolitan*, however, the heroic story of where Todd had traveled and what he had seen was all the more heroic because of its scientific accomplishments, which were encapsulated in the photographs that supposedly showed canals on Mars. As the *Cosmopolitan* editor put it to Lowell, the scientific aspects of the travel narrative "would appeal to the million readers of the *Cosmopolitan* and help toward a greater popularizing of the wonders of astronomy." Furthermore, he claimed, the report would be worthless and unappealing to readers if it had already appeared in *Century*, meaning he would have "no alternative, therefore, except to abandon the project which, as it lay very near my heart, I am sorry to do." Both editors threatened Lowell with this bluff, claiming that they would not publish anything at all if they did not have exclusive rights to first publication of the full expedition narrative.[51]

In the end, Lowell brokered a deal with *Cosmopolitan*, promising permission to publish one or two Andes photographs on the condition that they not appear for a month after the *Century* article was to be published.[52] Although the arrangement and condition were apparently grudgingly accepted, Lowell's publisher still worried that his prized exclusive could slip away. All of the wrangling over publication rights had been conducted by mail while Todd and Slipher were still en route to the United States. As such, it could not be expected that Slipher, as the person responsible for transporting the photographic plates, had any understanding of the feverish urgency with which publication arrangements had been crafted. Two days before Slipher was due to arrive in New York, the *Century* editor wrote to Lowell in a panic requesting the name of his steamer and details about his arrival:

> If Mr. Slipher should innocently show one of those photographs to a *World*,
> *Journal* or *Herald* reporter the chances are that it would appear the next day
> in one of these papers. In such matters the reporters of the yellow press are
> educated to get desirable material at any cost and in any way; so we shall have
> to outgeneral them at the dock. A telegram from you to Mr. Slipher would
> perhaps catch him at quarantine. You see that, having undertaken to give the
> first report of this expedition, we do not want to be defeated.[53]

This suggestion apparently did the trick, as Lowell's telegram alerted Slipher to the importance of protecting the plates' secrecy. *Century* reported calmly on October 5 that "Mr. Slipher is here and has received your telegram and is working in entire cooperation with us."[54]

These various competitions over the expedition chronicles—between publishers who wanted first-publication rights and between astronomers who wanted the prestige of association with a successful expedition—underline the ways that travel *writing* was as much a component of scientific legitimacy as the undertaking of scientific travel itself. The second half of the nineteenth century had been marked by the European nations' shift from maritime to interior exploration of potential colonial spheres along with a rising American interest in the resource potential of its Latin American and Pacific neighbors.[55] With this shift came an explosion of travel writing from various explorers competing for popular and official influence. Beginning with Alexander von Humboldt, the German natural scientist whose work launched the discipline of geography, Driver argues, "The authority of the explorer, in fact, depended substantially on the writing of a narrative of travel, either first or second hand."[56] Descriptive writing about sights and experiences in foreign landscapes, however, was not a specialized skill in the same way as surveying, mapping, or specimen collection. By writing narratives of their passage through foreign landscapes, with commentary on what they saw and where they went, even tourists, diplomats, or businessmen could cultivate an aura of scientific legitimacy linked to the enthusiasm for exploration.

The genre of the travel narrative, or travelogue, relied largely on the authority of eyewitness views and the sensation of heroic adventure to captivate broad popular audiences. It was fundamentally a tool of the geographer's gaze, based on direct visual observations made in specific places. The legitimacy of these accounts stemmed mainly from the fact that their authors had been in a place that the reader had never seen. By simply looking around and collecting images of the landscape—in textual descriptions, sketches, and photographs—the traveler produced geographic knowledge. Although nineteenth-century travel narratives were long taken at face value as geographical texts, much recent scholarship has pointed out that the genre was largely complicit in the imperial project, typically portraying foreign (especially tropical) lands as empty or welcoming to the European colonist or mercantilist.[57] This recent critical attention to a genre formerly considered innocent or harmless has revealed it as a complex locus of representation, where Self and Other were coproduced in ways that alternately reinforced and destabilized traditional hierarchies of race, ethnicity, and gender.[58]

In adopting the powerful voice of the traveler in their writing, even astronomers—who ostensibly focused their gaze on the heavens rather than on the landscapes through which they traveled—participated in the broader hegemonic conventions of the genre. In Todd's descriptions of South America's Pacific coast, for instance, he spent rather little time discussing the many cities he saw before choosing to disembark at Iquique, Chile, "in the midst of [a] forbidding waste" on the margins of the Atacama Desert. In Todd's words, "The region is an utter desert; the moon itself could not reveal greater barrenness–not a tree or a flower or a blade of grass for miles, not even moss or lichens."[59] Likewise, Todd barely mentioned the Peruvians or Chileans he met, yet he thanked by name the managers of the Western nitrate mining company who agreed to let him set up his telescope near their operation.[60] Todd's wife, Mabel, characterized the native Chileans as "an interested throng of dwellers" whose "undue curiosity" led them to gather outside a high fence around the telescope, where they stood "gazing with most amiable sentiments at the mysterious activities within, their wide sombreros, gay ponchos, silver spurs, adding picturesqueness to the scene."[61] The Todds' writing—for this expedition as well as for previous solar eclipse expeditions to Japan, Tripoli, and many other locations—thus implicitly celebrated the inroads of Western capitalism and science into supposedly barren and enticingly exotic landscapes, where native peoples were viewed as merely part of the scene. Pang has shown that astronomical expeditions to observe eclipses fundamentally relied on the infrastructural "tools of empire" to support their activities, so perhaps it should not be surprising that Mars observers' narratives also reflected this dependency and appreciation to some implicit extent.[62]

Astronomers' expedition reports and narratives, then, should be read as intricate legitimacy maneuvers that were constrained by convention and by audience expectation in ways that deeply affected their scientific claims about Mars. First and foremost, travel narratives required astronomers to represent themselves as hardy masculine heroes. Like Bailey's and Pickering's mountain-climbing exploits in Peru, Todd's expedition was presented as a story of bravery and heroism. Although Todd traveled in relative comfort and with the benefit of numerous diplomatic contacts, he characterized his South American landing in Iquique in dramatic terms: "abreast of the extensive desert wilds of Tarapacá, about as far as seemed safe . . . I resolved to land the telescope and risk the fate of the expedition."[63] This overblown statement about "risking the fate" of the expedition was calculated to satisfy readers whose expectations had been heightened by sensational early announcements that the expedition would "proceed overland from Colon to Lima, Peru [then] begin a rough journey up the mountain range," eventually establishing the telescope "18,000 feet high in the Andes, where the atmosphere is noted for its clearness and steadiness."[64] Instead, Todd's party made most of the journey by boat, landing in a city where the American vice-consul immediately introduced Todd to officials from the nitrate company who granted full access to their railway and daily train service. Within four days, the telescope had been transported and erected on a concrete tennis court (!) in front of the manager's house at the nitrate workers' settlement (causing what Todd jokingly characterized as "a somewhat disastrous interruption of the happiness of tennis experts"). Despite the apparent ease with which all of this was accomplished, Todd felt obliged to note the *potential* for difficulty by diverting readers' attention to the exotic landscape in which he bravely pursued his science: "around the whole settlement stretched the solemn, brown, impressive pampa, undulating to the great mountain border, the Andes, its peaks here and there snow-capped, lofty, and magnificent."[65]

This diversion apparently worked, as newspaper accounts of Todd's journey and work focused on the forbidding and mountainous landscape. Despite the fact that most of Todd's work was conducted in Alianza—at 3,000 feet in elevation, well below the Andean peaks to the east—for instance, a *New*

York Times feature story repeated the original predictions that Todd's observing station would be established at 16,000–18,000 feet above sea level. The same article then quoted Todd extensively on the subject of mountain-sickness at high altitudes, quoting comments he had made in reference to a limited side expedition to higher elevations in the Andes.[66] Newspapers also seized on Todd's remarks that the expedition party feared its telescope might be damaged by an earthquake or a tidal wave (at Iquique), thus focusing on landscape difficulties that heightened the heroism involved in carrying out Mars observations. Todd himself embraced the dramatic voice that his audience expected of a travelogue, casting himself as a heroic figure who became all the more heroic for sensational scientific findings that confirmed the existence of canals on Mars. This followed an established model of narration that Todd had learned from the solar-eclipse expedition circuit. As Pang noted of the many astronomers who wrote dramatic stories of their eclipse-related travels: "This was a group that knew what would sell, and narrow escapes from Arab porters and threatening clouds were highlighted to satisfy the needs of commerce as much as science."[67]

At the same time, Todd's use of the travel narrative deeply affected the methods he used to perform his science. In writing as travelers, astronomers essentially compelled themselves to use the geographer's gaze as a methodological tool of both vision and interpretation. The heroic story of moving through the foreign landscape was essentially a story about moving into position for a direct view on Mars. Without having undertaken the travel, Todd suggested, he would not have been able to cast a direct gaze on Mars or to legitimately interpret the meanings of what he saw through the telescope or in the photographic plates. This interpretation, of course, was based on the analogical reasoning of the geographer's gaze, as seen in Todd's repeated use of the Chilean pampa landscape to explain his understanding of Mars:

> Old earth again furnishes a ready clue to the mystery. . . . The more I visit arid regions of the earth, and observe the devices of desert-dwellers to coax the growth of even the sparsest vegetation, the more the truth of Lowell's theory of the Martian canals impresses itself upon me. During this last summer, in the desert of Tarapacá and in similar wastes of Peru, I saw vast areas, or oases, saved from engirdling sands by just a little water—water not in great gulfs or rivers or lakes, but a tiny rivulet merely, systematically diverted from its course again and again, with the parched soil divided and subdivided in geometric figures till nothing was left of the original stream but an infinitude of trickles. But as we approached these oases of the Chilean mountains, or receded from

them, they seemed one vast and consecutive mass of vegetation, much darker than the desert around. Imagine yourself suspended high above such terrestrial sands, as in a balloon, only hundreds or thousands of miles away, and the likeness of Mars to the earth and the earth to Mars would be compelling.[68]

In this text, Todd's commentary reveals a deep intertwining of physical movement, visual observation, and the seemingly inevitable likeness of Mars to Earth. In the same article, Todd explicitly noted numerous terrestrial analogies, saying for example that conditions on Mars "cannot be very dissimilar to the lofty mountain heights of the Andes, where I found man and many types of animal and vegetable life enjoying a useful existence, in spite of what Professor Lowell excellently calls 'pronounced inhospitality of environment.'" In addition to these descriptive and interpretive analogies, Todd also summoned the Andean high desert landscape in other creative bids for legitimacy, claiming that the nonhuman landscape almost cried out in support of Todd's (and Lowell's) interpretation of Mars: "Nearly everybody who went to the eyepiece saw canals; and once I fancied I heard even the bats, as they winged their flight down the pampa, crying, 'Canali, canali, canali!'"[69] Todd's specific claims and methods of legitimacy cultivation thus reflected the importance of using the geographer's gaze as a standard element of travel writing.

Another important effect of the travel narrative on Mars scientists' legitimacy was the destabilization of established disciplinary hierarchies based on professional status and reputation. As mentioned in the previous section, travel and travel writing have been shown to have a fluidizing effect on traditional identities and hierarchies. Blunt shows, for example, that British travelers were defined more by their imperial (racial) identities while traveling than by the domestic (gendered) identities that defined them at home, thus requiring female British travelers to navigate a complex transition of identity that involved the adoption of different voices, depending on where they were and for whom they were writing.[70] For the scientific disciplines specifically, this fluidization allowed those scientists (or even nonscientists) with lower professional standing to mobilize more authoritative identities, often making an outsize impact in the popular press by virtue of their travel writing. As described in the preceding section, Pickering's science was seen as legitimate largely because he had gone heroically into the field to gather his data, producing knowledge so apparently groundbreaking that it simply could not be reported in the standard venues of the academic journals. Despite sometimes vigorous skepticism (or even outright protest) toward travelers' scientific claims from within the professional societies, the stigma of the "armchair" scientist inhibited the ability of even respected elders to pass judgment on the upstart scientific travelers, at least in the popular press.[71] The traveling astronomer thus gained the upper hand through access to different audiences that perceived scientific legitimacy less rigidly than the professional communities.

Both Todd and Lowell used the 1907 South American expedition to capitalize on this disruption and enhance their popular reputations. As noted, Todd was a veteran astronomical traveler and had long used the genre of the travel narrative to enhance his popular standing. In 1907, however, he found himself in competition with Lowell for control of the legitimacy that accompanied the expedition. Even though Lowell himself did not travel to the Andes, he was very savvy in controlling the narratives about the expedition. He was the one who sent reports to the newspapers, and he ensured regular mention was made of the expedition in any reports on his own continuing observations from Flagstaff. The eventual struggle between Lowell's publisher and Todd's publisher over rights to the expedition reports and photographs reflects the attempt by Lowell to claim and divert some of the expedition's legitimacy to his own personal reputation. Did it work? If the grumblings of establishment astronomers are any indication, both Lowell and Todd received major boosts to their reputations. Personal correspondence between Campbell (director of the Lick Observatory) and Hale (formerly director of Yerkes Observatory, then director at Mount Wilson) reveals a sense of futility and exasperation at the effect travel narratives had on popular audiences:

> You have of course noticed that Lowell, the past year or two, has been making much ado in public, and in many matters quite unprofessionally. I have occasionally thought of putting my finger publicly on the weak points, but have serious doubts as to the usefulness of such an unpleasant undertaking. I do not believe that either his photographs or Todd's record any markings on Mars that have not been conceded to exist by experienced observers of the planet for twenty-five years past; and which can be seen with a six-inch telescope better than they have been photographed. . . . I think Lowell and Todd are going to be a trial to sane astronomers. . . . My question is just how far they should be allowed to go before somebody steps on their rope.[72]

Despite significant concerns regarding the accuracy, legitimacy, and originality of Lowell's and Todd's science, even the most highly respected astronomers in the United States felt somewhat powerless to do anything about it.

Campbell's comments reveal an important distinction: between popular legitimacy (with broad, literate audiences) and professional legitimacy (with scientific practitioners). Travel and travel writing had first contributed strongly to both kinds of legitimacy, but Todd's 1907 expedition to South America took place during a time of transition that saw the criteria for professional legitimacy revised. As more and more Europeans (including women) began traveling and recording their experiences for home audiences throughout the nineteenth century, the professional societies found themselves facing the uncomfortable reality that fame and legitimacy were suddenly available to a large group with no scientific training, pedigree, or practical ability. To use a conflict from the discipline of geography as an example, the sensational travel narratives written by the newspaperman Henry Morton Stanley of his African explorations in the 1870s were first met with disbelief and outrage at the Royal Geographical Society (RGS), whose fellows considered him an impostor, a rogue, and certainly not a legitimate geographer. But as Stanley grew increasingly popular with wider audiences, the RGS was prompted to swallow its pride and award him one of the society's medals in a "face-saving gesture" that was "much against the wishes of many fellows."[73]

Conflicts like this one prompted the guardians of geographic knowledge to counteract the proliferation of travelogues in the publishing landscape by distinguishing between "scientific" exploration and "descriptive" geography. Those who hoped to gain scientific legitimacy for their fieldwork thus became obliged to carry measuring instruments, to keep quantitative field notes, and to write of their experiences in impassive and detached tones. To a large degree, scientific exploration became a masculine domain while nonscientific travel became somewhat feminized.[74] The professionalization of geography and other field sciences deepened (and depended on) this distinction, restoring the boundaries of scientific legitimacy toward the end of the nineteenth century.[75] In this new era, it was dangerous to attempt an appeal to both scientific and popular audiences, for being labeled a popularizer could be a death-knell for a promising scientific career.

Given the development of this new distinction, it is surprising not that it dogged both Todd and Lowell during the 1907 expedition but rather that both of them managed to rise above it and produce influential scientific results nonetheless. Lowell, for example, managed to win support from numerous scientists in other disciplines despite his sensational publishing strategies and the frequent characterization of him as a popularizer. Even though he was never fully accepted by mainstream astronomers during his twenty-year career, his theories about Mars actively drove the development and direction of Mars science. Todd, on the other hand, had established a fairly secure professional reputation on the basis of regular eclipse expeditions that saw him transport massive scientific equipment into the field and then return with quantitative data. In 1907, however, his reliance on dramatic travelogue-style writing went well beyond what would have been considered acceptable for a legitimate scientist.

Todd even went so far as to illustrate one of his primary articles with a sketch of Mars his wife had completed during the expedition (see fig. 4.4). Although it was not unusual for a woman to accompany male astronomers on an expedition, it was in fact extraordinary that a drawing by a woman was published as scientific evidence of an astronomical finding. Within astronomical work, women were physically excluded from British astronomical expeditions and textually removed from reports of the American expeditions that allowed them to travel and even relied heavily on their skills in organizing and maintaining a camp.[76] In general, women who aspired to scientific work while traveling typically found their activities sharply circumscribed by gendered expectations that limited their contributions to descriptive sciences like botany. In writing about their journeys for popular audiences, women adopted what were traditionally considered male roles—gazing upon and describing a landscape from the vantage point of the authoritative subject—and "became increasingly able to share in the authority of male colonizers." At the same time, however, women were expected to gaze differently and write about their experiences in different ways than their male counterparts. For women, theirs largely became the role of the subjective and personal narrator: one who could gaze upon and take in a natural or cultural scene—even in heroic ways that transgressed expectations of domesticity—but had little control over that scene. The inclusion of Mabel's sketch thus shows that Todd was little troubled by the need to present himself as a hard-core scientific explorer. Instead, he embraced the mode of descriptive geography in his reports on the South American expedition.[77]

This raises several interesting questions. Why didn't Todd's embrace of descriptive rather than scientific travel (and travel writing) taint Todd and Lowell in ways that automatically rendered their scientific claims illegitimate? How was Mabel's sketch creditably published in a leading magazine? Why couldn't Campbell and Hale give public voice to their complaints? For astronomy, it seems, descriptive geography was not necessarily inferior to scientific exploration. The 1907 expedition was represented as both scientific exploration and as adventurous travel, with few detrimental effects. Once again, the fundamental nature of astronomy—as a nonfield science that was occasionally practiced in the field—seems to have insulated the

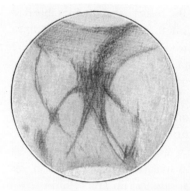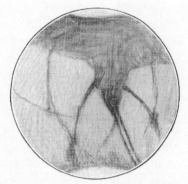

. DRAWINGS OF MARS AS SEEN THROUGH THE GREAT TELESCOPE LAST JULY
The above drawings were made by Mabel Loomis Todd at Alianza and were sent to the Cosmopolitan by Professor Todd. They show how plainly the double canals appeared to the observer
through the clear atmosphere of the higher Andes.

Figure 4.4. Sketches of Mars by Mabel Loomis Todd, 1908. These sketches appeared in
American astronomer David Todd's first major publication about the results of the 1907
Lowell expedition to the Andes, in *Cosmopolitan Magazine* ("Professor Todd's Own Story," on
p. 348). As the caption notes, the sketches were completed by Todd's wife, a nonastronomer.
They were among only a few illustrations that accompanied the article, with the others
being a photographic contact sheet, a composite Mercator map of Mars, and two
photographs showing the Amherst telescope mounted in Alianza, Chile.

Mars astronomers from the travel-versus-exploration distinction that afflicted other field sciences. Gazing from a foreign landscape rather than on
it, and interpreting results via comparison to that same landscape, was a
new kind of gazing that operated under different legitimacy rules. The next
section examines this phenomenon from the reverse standpoint by analyzing an astronomical expedition that was designed as a purely scientific
exploration, without any pretense of travel writing or popularization, and
which failed to definitively trump the Lowell-Todd expedition despite producing contradictory results.

The New Explorers: Wallace Campbell on Mount Whitney

In answer to the popular furor over Martian canals that Lowell and Todd
stoked with their 1907 expedition, Campbell plotted his own expedition: to
the summit of California's Mount Whitney in 1909.[78] Just like the Lowell-
funded expedition, Campbell's was a carefully planned endeavor meant to
settle the life-on-Mars debate by cultivating unimpeachable legitimacy for
his scientific claims. But Campbell made a point of eschewing the popular
press and maintaining the aura of the scientific explorer in his writings. He

certainly lapsed into heroic narrative and relied on a geographer's gaze in his writings about the expedition, but his approach was markedly different from Lowell's and Todd's. He published only in scientific journals and diligently tried to keep the expedition a secret so as to prevent any sensational reports in the newspapers.[79] Campbell's strategy met with mixed results, indicating that differing levels of legitimacy were produced through the use of two competing personae: the heroic gazing traveler and the rational scientist-explorer.

In August 1909 Campbell, then director of the Lick Observatory, left his home near the summit of Mount Hamilton in California's Diablo Mountains and traveled with a small expedition party to Mount Whitney in the Sierra Nevada Range. His purpose was to observe the planet Mars with spectroscopic instruments that would allow him to settle a simmering debate over whether the Martian atmosphere contained any measurable amount of water vapor and, thus, whether the red planet might be habitable. Spectroscopes, which record the wavelengths of light reflected or emitted from a celestial body, had been used to assess the Martian atmosphere since the 1860s.[80] Because different chemical elements produce different spectral signatures in reflected solar light, an observer on Earth can determine and characterize—by virtue of the exaggeration or dullness of various wavelengths captured by the spectroscope—the existence and composition of an atmosphere. Although spectroscopic data collected since the 1860s had indicated the presence of water vapor in the atmosphere of Mars, Campbell had disputed these findings beginning in 1894, when his own spectroscopic work showed virtually no detectable water vapor.

In 1908 Lowell published a table of spectroscopic results by his staff astronomer V. M. Slipher, offering it as proof of an aqueous atmosphere on Mars.[81] Campbell, who had long wanted to repeat his 1894 results under excellent observing conditions, took Slipher's work as his impetus and set out to silence the inhabited-Mars proponents. To achieve his goal, Campbell had determined that measurements were needed from the "highest point of land in the United States," where the density of Earth's own atmosphere would be lowest and therefore least disruptive to the very sensitive photographic processes required to assess the composition of the red planet's atmosphere.[82] From Mount Whitney's 14,000-foot peak, highest within what was then the United States, Campbell intended to compare the spectrum of Mars with the spectrum of the moon, which was widely acknowledged to have no moisture or atmosphere. Because the effects of Earth's own atmosphere must also be accounted for in the wavelength disturbances recorded by the spectroscope, spectroscopic studies often involve the comparison of two or more celestial bodies, thus allowing for any common anomalies to be discounted as due to Earth's atmosphere. Campbell set the dates for his Mount Whitney expedition in order to maximize the ease of turning the telescope from Mars to the moon quickly, while they were both at the same height in the same part of the sky.

Starting from the village of Lone Pine on August 25, 1909, Campbell and his group traveled by carriage and horseback up the slopes of Mount Whitney to a base camp at 10,300 feet (fig. 4.5). After two days spent adjusting to the effects of altitude, they continued their ascent on pack animals through threatening weather and snow showers. After a difficult final ascent, they reached the shelter at the 14,000-foot summit just as a lightning storm began. After setting up their temporary observing station and adjusting instruments in continuing bad weather, the astronomers welcomed a clearing of the skies that provided atmospheric conditions that Campbell described as being "as perfect for our purposes as could be wished" for the next two nights. During this time, Campbell exposed numerous spectroscopic plates of the moon and Mars, while other members of his party took meticulous measurements of the atmospheric conditions atop Mount Whitney. Although clouds and snow prevented completion of the planned third night of observations, Campbell was satisfied that he had gathered good data under excellent conditions. On his third and final day of observations, in fact, Campbell reported that "the sky was absolutely clear; the wind was from the fair-weather quarter; the humidity was low; and the sky was remarkably blue. On occulting the Sun behind the roof of the shelter one could look up to the very edge of the Sun with no recognizable decrease in blueness. I had never seen so pure a sky before."[83] Throughout his report from this expedition, Campbell emphasized the purity of the view from Mount Whitney and presented it as an ideal site from which to gaze upon Mars.

During this expedition, Campbell captured what he considered to be conclusive proof that the Martian atmosphere had no detectable moisture. Campbell's plates showed the spectra of the moon and Mars to be identical. Because they were intentionally observed at virtually the same time, "at equal altitudes, in quick succession, with the atmospheric conditions substantially unchanged between observations," Campbell argued that Mars must have an extremely thin atmosphere, or none at all, just like the moon. Publishing his findings in a widely circulated *Lick Observatory Bulletin*, he was careful not to rule out the possibility of a Martian atmosphere entirely. But if Mars did have an atmosphere, Campbell argued, his findings proved that it could not "exceed or equal the quantity of terrestrial vapor above the

FIGURE 145.—MOUNTAIN CAMP, MT. WHITNEY CALIFORNIA
(12,000 feet).

Figure 4.5. Sketch of astronomers' mountain camp on
California's Mount Whitney, 1896. This image, which appeared
as the frontispiece of the October 1896 number of *Publications of
the Astronomical Society of the Pacific*, shows a high-altitude base
camp on the slopes of Mount Whitney. Campbell's description
of his party's climb to Mount Whitney's summit focused on both
positive and negative effects of high altitude. This image also
appeared unchanged in Holden's *Mountain Observatories
in America and Europe* in 1896.

same area of Mt. Whitney."[84] In other words, a Martian atmosphere dense
enough to support extensive life was extremely unlikely.

Campbell's findings were welcomed as conclusive by professional American astronomers who were hoping to dent Lowell's credibility. In reporting that
the Martian atmosphere contained less water vapor than could be perceived
by modern instruments, Campbell had "put the burden of proof" on those,
like Lowell, who relied on a wet Martian atmosphere as the basis of their
arguments regarding Martian canals and inhabitants. Campbell felt that his
expedition had settled the issue by taking unimpeachable data from a much
higher altitude than "at all the observatories where the Martian spectrum
had previously been investigated."[85]

Mainstream audiences, however, were not so convinced, and Campbell's
opponents at the Lowell Observatory in Flagstaff continued to insist on the
existence of both water and life on Mars. As a result, Campbell was still defending his findings more than twenty years later.[86] Why didn't Campbell's
mountaintop science produce results more widely accepted as conclusive?
The sites of science were important parts of the legitimacy constellations
surrounding Mars science. Yet Mount Whitney—a remarkable site that was
meticulously selected, heroically ascended, and scientifically conquered—
failed to produce finality beyond the community of professional astronomers. Part of the explanation for why Campbell never achieved that finality
lies in the way he represented himself and his activities upon the imposing
summit. To be sure, Campbell engaged in heroic self-representations that
emphasized the importance of the mountain as a sublime and challenging
site. But Campbell fundamentally rejected the travelogue mode of representation, eschewing a heroic-traveler persona and adopting an explorer-scientist
persona instead.

Campbell was by 1909 an accomplished astronomer, spectroscopic observer, observatory director, and expedition leader. He led the Lick Observatory
in the same exacting way that he had conducted numerous successful eclipse
expeditions around the globe.[87] From Jeur, India, in 1898 to Thomaston,
Georgia, in 1900 to Alhama, Spain, in 1905 and Flint Island (in the Pacific)
in 1908, "the Lick expeditions were famous for their detailed advance planning and testing, transport of delicate scientific equipment to a remote destination unharmed, and days of practice so that at the brief moment of
eclipse all the instruments were operated flawlessly." Although the planned
expedition to Mount Whitney's summit was designed to maximize observations over several nights—a considerably longer window of opportunity
than the mere minutes available for solar eclipse observation—the stakes
were similarly high. Campbell clearly "regarded the search for water vapour

in the atmosphere of Mars as an important scientific problem, but it is also clear that he wanted to prove Lowell wrong and take him down in public estimation."[88] In preparation for this important task, Campbell selected Mount Whitney well in advance, made a reconnaissance trip in 1908, commissioned the building of a shelter on its summit, and secured both funding and a crew to make the mission possible.

Campbell's attention to detail and his focus on instrumentation and data collection clearly enhanced the confidence with which he presented results to the scientific community. But he also made effective use of other representational strategies. At one level, Campbell painted himself fairly successfully as a heroic traveler, using physical movement as the basis for settling a legitimacy dispute. He described in his official report a very difficult ascent of the mountain and a painstaking setup of his scientific instruments, made especially arduous by harsh weather. The expedition party's arrival at the mountain's peak, for instance, could hardly have been more dramatic, for just as "Director Abbot opened the door to receive us there were two violent discharges of lightning, near enough to be felt by most members of the party." Campbell celebrated the remoteness and uniqueness implied by these difficulties while also celebrating sublime mountaintop conditions, as in his rapturous claim that he "had never seen so pure a sky before."[89] These powerful representations of the expedition were critical to the scientific legitimacy of its results, which might otherwise have been seen as extremely limited and inconclusive, given that Campbell observed Mars with the spectroscope on only two nights and reported stormy weather both before and after. By emphasizing that he had overcome physical challenge to achieve a direct gaze on Mars, Campbell mobilized the typical trump cards of fieldwork representation: strenuous physical activity and masculine heroism.

All the same, however, Campbell distanced himself from popular legitimacy in his bid to preserve the rational stance of the explorer-scientist: one who is not caught up in the physicality or emotion of the moment but who relies on instruments to gaze and objective reason to interpret. When Campbell first began to investigate the possibility of an expedition to Mount Whitney's summit, for example, he couched his intentions in vague language, saying simply that he was interested in a "certain astronomical problem demanding high altitude."[90] He asked associates not to comment on the expedition and sent no cables to the press. Osterbruck explains this intentional vagueness by characterizing Campbell as "a scientist of the old school [who] abhorred any publicity before the results were completed and published in a respectable scientific journal."[91] I would argue, however, that Campbell was operating less in an "old school" context than he was responding to very new conditions in the creation of scientific legitimacy. Where scientists had earlier been able to capitalize on the masculine heroism of the traveling adventurer in their representations, the professionalization projects of the late nineteenth and earlier twentieth century (in astronomy and other disciplines) closed this door to those who aspired to professional status. To avoid charges of amateurism, such as those that continually afflicted Lowell, Campbell was limited in how he could represent himself. Cabling the press with news of his dramatic exploits would have undermined his professional standing and lessened the impact of his findings on the professional community. By avoiding all but the most limited representations of himself as a heroic explorer, however, Campbell missed an opportunity to shift the popular debate on Mars.

Conclusions

Given that Campbell went on to become president of the University of California and then president of the National Academy of Sciences, his restrained representations of the Mount Whitney expedition seem prudent in hindsight.[92] Yet the fact of his inability to truly conclude the Mars water-vapor debate in 1909 speaks to the importance and uniqueness of the moment in which the Mars debates unfolded. During the window in which scientific professionalization was actively pursued yet not fully established, there was a transitional period in which professional (and aspiring professional) astronomers were actually less influential than amateurs or popularizers on some topics, given their need to restrain their self-representations and their outreach to popular audiences.

For the discipline of astronomy during the years of the Mars canal sensation, this paradoxical situation was further complicated by the physical movement of observatories to Western mountains as part of the professionalization project. As scientific astronomers went out into "the field," popular audiences began to connect their views of astronomy with existing popular enthusiasm for field science, adventurous travel, and even recreational sport in remote mountains and high altitudes.[93] To some extent, professional astronomers were able to capitalize on this enthusiasm through the masculine persona of the explorer-scientist. Claiming legitimacy as observers on the basis of their remoteness and emphasizing the necessity of secluding themselves in distant and pristine landscapes, astronomers embraced the new persona. Holden's book on mountain observatories, for instance, included images of astronomers who looked every bit as hardy as the seasoned polar and glacial explorers then making headlines throughout Europe and North America (see

FIGURE 1:—ON THE WAY TO THE MONT-BLANC OBSERVATORY.

Figure 4.6. Illustration of astronomical party's ascent of Mont Blanc, 1896. In this image, which appeared in Holden's *Mountain Observatories in America and Europe* (facing p. 22), astronomers make their way through an ominous vertical landscape toward Europe's highest-altitude observatory atop Mont Blanc.

figs. 4.6 and 4.7). Perhaps needless to say, Holden's volume contained numerous such depictions of miniscule figures in ominous vertical landscapes but not one image of a passive astronomer seated at a telescope. Advocates of the inhabited-Mars theory, particularly, portrayed their scientific activities as rigorous, strenuous, and adventurous in order to assert superiority over their critics. This had a significant positive impact on the credibility of such astronomers with nonspecialist audiences.

But there were limits to how far professional astronomers could go with these representations before their legitimacy within the disciplinary community was compromised. Those astronomers like Lowell, who were willing to forego full professional membership, were able to embrace self-representation as travelers, field scientists, and gazers, thus exciting popular audiences' imaginations to such an extent that the professional community lost control of the scientific debates. In the process, the astronomer-geographers deeply impacted the nature of scientific knowledge production, affecting even those who disdained their methods and influence.

Throughout the professional, semiprofessional, and amateur ranks, the landscape gaze became the dominant means of understanding Mars.

FIGURE 16:—CHIMBORAZO FROM A POINT 17,450 FEET ABOVE SEA.

Figure 4.7. Illustration of astronomical party's ascent of Mount Chimborazo, 1896. In this image, which appeared in Holden's *Mountain Observatories in America and Europe* (facing p. 43), astronomers labor on the slopes of South America's Mount Chimborazo.

Claiming to see the planet with a direct, unimpeded view and then making sense of that view through intuition and analogy—either to general topographical features and geographical processes, or to specific landscapes, cities, and countries—was a rhetorical staple not only of the inhabited-Mars proponents, but also of their critics. This unexpected (and perhaps, in some cases, unconscious) devotion to a method derived from another discipline—geography—therefore invites reflection both on the nature of Mars science and on the nature of the gaze itself. As Rose pointed out so incisively, the geographical way of seeing is by its very nature masculinist (given the mobility it requires), imperialist (given the control or possession it implies), and remarkably powerful.[94] Yet it is also inherently unstable. The seductions of a landscape open to penetrating gaze stir geographers and travelers to admit taking pleasure and awe in the act of looking, possessing, and knowing.[95] While such admissions reinforce the power of the gaze, they also call into question the objectivity of the gazer. In the history of geography as a discipline, this ambivalence produced a methodological and representational hybridity, where geographers tried to temper claims of a direct gaze with representations of analytic distance, lest one be labeled little more than a pleasure-seeker. Outram suggests that the rise of museums and the proliferation of exhibitions in the late nineteenth century reflected this need to overcome the ambivalence of the gaze and the conflicts of legitimacy it engendered.[96] In a museum or exhibition, the geographer's gaze can be direct and pleasurable while at the same time maintaining an objective distance for both the producers and consumers of the exhibits. The geographic knowledge produced in such spaces—where vision is privileged over other senses—is therefore untainted by the physical seductions associated with being in the field, even though it remains based on a fundamentally masculine (and imperial) method of knowledge production: the gaze.

The astronomer studying Mars seems to have found another way of resolving the same conflict, thus faring better than the field geographer who could be accused of taking too much pleasure in the gazing task to produce rational and believable knowledge. Despite taking pleasure in their direct view of the Martian surface and reacting with awe to the notion of an advanced civilization revealed in the red desert landscape, astronomers were largely protected from any loss of objectivity by the simple fact of physical distance. Lowell could simultaneously claim to be "traveling" to Mars in spirit to gaze upon its landscape and to be perfectly objective and impervious to the follies and seductions of the polar explorers. Where geographers were careful to guard against accusations that they were mere travelers, it was hard to accuse any astronomer of being a "traveler," no matter how complete the loss of logical reasoning that seemed to accompany the very real seductions of the Martian landscape. There were indeed claims from all sides that various opposing astronomers had "lost objectivity" or lost sight of logic or reason, but none of these accusations were associated with the gaze as methodology. Astronomers were thus free to devote themselves to claims of direct and pleasurable gazing and to cultivate representations of themselves as field explorers and travelers without unduly risking their popular reputations as objective scientists.

Where their professional reputations were concerned, on the other hand, astronomers faced some of the same concerns as the scientific geographers. The visual methodology of the gaze was acceptable, but the interpretive methodologies of terrestrial analogy were somewhat more dangerous, given the seductions they implied. When Todd claimed to hear the Chilean bats crying "canali, canali" in their flights overhead, he cultivated enduring legitimacy among popular audiences while simultaneously deeply alienating himself from the community of professional astronomers. Todd's position of legitimacy between these two audiences is of course quite similar to the liminal position of the lab/field border zone (discussed in chapter 3). Astronomy managed to conflate the distinction between lab and field by emphasizing that elements of the observatory site allowed for both objectivity and direct contact. In their self-representations, some Mars scientists similarly managed to present their use of the geographical gaze as both direct and distanced. Pickering and Todd conducted very real fieldwork: they were present in a landscape that altered their perspective, and they admitted openness to the pleasures and awe of looking. Paradoxically, however, they were never actually present in the landscape that they were studying. So there never was any conflict between contact and distance—they claimed to be able to achieve both.

The inherent instabilities of the gaze (produced by the conflict between direct perception and objective distance as legitimate ways of knowing a landscape) were thus sidestepped by astronomers in ways that geographers themselves could never manage. The fact that astronomers conducted fieldwork, used a rhetoric of direct contact, and relied on the interpretive processes of the geographical gaze yet were never actually present in the landscapes they studied allows for a different reading of the construction of a Martian landscape. If we follow Cosgrove and Daniels in characterizing landscape as a cultural image, we can see that for the case of Mars, the observed landscape was truly nothing more than a way of seeing.[97] The heroic (masculine) physical acts involved in pursuing this vision were easily conflated with travel narrative and (feminized) descriptive geography. The mental acts involved in constructing this vision expressed a distinct and traceable relationship between nature and society. The next chapter will explore more fully what the constructed Martian landscape reveals about this nature/society relationship.

FIVE

Placing the Red Planet: Meanings in the Martian Landscape

The facts are for the most part now established. The question turns entirely upon their interpretation.

—American sociologist Lester Frank Ward (1907)

If we accept the evidence of the foregoing chapters—that Mars astronomers and their audiences relied on geographical methods of investigation, interpretation, representation, and legitimation—we can examine Mars science and sensation as geographical science and sensation. In the September 1896 issue of *Harper's*, for example, the leading American magazine's editor, Charles Dudley Warner, focused his monthly "Editor's Study" on the meanings of recent Martian science. Reviewing the recent history of research into the planet's characteristics, he expressed great confidence in the cartographic work of the past two decades:

> We have ascertained his comparative age, his size, his weight, his common habits as a planet, his climates, we have given him a geography, mapped out his surface, and supplied every portion of him with names, un-Earthy and undescriptive, and, in short, done everything that enterprising speculators could do except to stake him out in quarter sections for pre-emption. We talk about his seas, his continents, his islands, his capes, his rivers, his canals, his oases, his deserts, his hills and valleys, his climatic zones.[1]

He specifically lauded Lowell's 1894 book *Mars* for its "excellent maps of this planet, which I have studied with an intense desire to make them as intelligible to me as those the old geographers made of Central Africa."[2] From this opening acknowledgment of the geographical inflections of Mars science, Warner then proceeded to a detailed rumination on the probable nature of Martian inhabitants:

> [T]he inhabitants are making a fight for existence which shows that they have attained a high development in the scale of being, and may have got rid of politics altogether. In fact, it is suggested that such a vast and regular system of irrigation, extending over the whole surface, would be inconsistent with hostile divisions into nations, with war, or even with "local option," or the fluctuations of popular suffrage. . . . For myself, looking at the matter from the Arizonian point of view, and accepting the theory of the very advanced development of life on Mars, I inclue to the opinion that the civilization has become refined instead of gross.[3]

Although he cautioned that these interpretations were nothing more than "wholly idle speculations," Warner obviously did not refrain from drawing such broad conclusions in the first place for his large audience of American readers. In that regard, he was no different from the editors and writers of other general-interest publications, who generally treated Mars as a topic of significant import. While the scientific journals and popular-science magazines focused mainly on questions of vision and certainty, the literary magazines, reviews, and newspapers were more likely to address philosophical, moral, and social meanings that they saw reflected in the inhabited Martian landscape.

In turning to these specific discourses, we must acknowledge that the broadest popular discussions of Martian landscapes often departed, in some cases significantly, from accepted scientific knowledge regarding actual conditions on the Martian surface. This departure, however, does not diminish the importance Mars carried during the decades of strongest belief in the Martian canals. If a "landscape's meanings draw on the cultural codes of the society for which it was made," those very cultural codes will be evident in the landscape itself.[4] This certainly holds for the case of Mars, even though the "making" of its landscape was a visual and textual, rather than material, process. By examining the scientifically created landscape of the red planet and its modification through popular interest and media sensation, we can begin to understand the deeper cultural meanings that drove its most heated debates. To a large extent, the debates over Mars were cultural negotiations about the nature of proper human–environment interaction.

This chapter examines the debated geographical meanings of Mars, putting them in context alongside then-emerging trends in geographical thought. This contextualization is conducted primarily through a focus on the competing arguments and contrasting intellectual positions of two prominent scientists: Percival Lowell and Alfred Russel Wallace. As high-profile interpreters of the latest Mars science, these public intellectuals reached large audiences and dominated Mars-related headlines with their mutual antagonism during the height of the canal debates. In what follows, I depart from traditional understandings of the Lowell-Wallace debate, which tend to cast it as a fundamental disagreement between secular Spencerian philosophy (as advocated by Lowell in his dogmatic support for the inhabited-Mars hypothesis) and natural theology (as espoused by Wallace in his steadfast refusal to accept a plurality of worlds). While not dismissing this intellectual context, I argue that the Lowell-Wallace debate and, indeed, the larger scientific and popular fascination with Martian geography, can better be understood in terms of emerging understandings of nature-society relationships within rapidly changing geopolitical contexts. Concerns about the global expansion of markets under imperial capitalism, the difficulties inherent in organization and control of laborers and landscapes, the deterioration of environments subject to resource extraction, and the specter of looming cultural conflict in an increasingly interconnected world were both reflected and modified within the debates over Mars.

The variously skeptical or enthusiastic responses of different scientists and audiences to the inhabited-Mars hypothesis were conditioned to a large extent by imperial contexts and the geographical sciences and theories that were then emerging to control and explain them. The common representation of Mars as a dying desert world dependent upon its irrigation infrastructure, for instance, was linked to a deterministic understanding of the relationship between culture and environment that played off stereotypes regarding arid, irrigated, and degraded landscapes. Looking at Lowell, Wallace, and their respective influences on audiences, debates, and intellectuals, then, helps illuminate both the Mars sensation and the nature of geographical knowledge production at the turn of the twentieth century.

Wallace and Lowell: Two Public Intellectuals

As discussed in preceding chapters, Percival Lowell was both reviled and praised in his own time, setting a pattern that has continued since. Historians have seen him alternately as a narcissist engaged in "self-indulgent fancy" or as a highly esteemed "sage" whose influence on Victorian intellectual currents was comparable to that of Carlyle or Ruskin.[5] The truth, of course, lies somewhere in between. As one of his obituaries put it, "While it is true that his astronomical theories scandalized staid old scientists, at the same time they attained a hold on the popular imagination which has never been loosened. . . . They did more to popularize the study of astronomy than all the college courses could have done in a hundred years." In fact, "more than a few" of the supposedly scandalized scientists were still fascinated by Lowell's theories into the 1950s, judging by research agendas and textbook treatments.[6]

While Lowell's attention to broad and general audiences marked him as a nonconformist among the professional ranks, his deliberate publication strategies allowed him to fulfill the role he conceived for himself as a public intellectual. Lowell consistently argued that it was the role of the scientist not only to report scientific findings, but also to make their meanings clear for general audiences. In the preface for his 1906 *Mars and Its Canals*, for instance, he justified his publication approach thus: "To set forth science in a popular, that is, in a generally understandable, form is as obligatory as to present it in a more technical manner. . . . The whole object of science is to synthesize, and so simplify." Under this mandate, Lowell found it not only acceptable but in fact necessary to discuss the ramifications of any theory that had not been disproved. His critics, on the other hand, argued that nothing should be discussed unless it was proven, as in Holden's stern evaluation of Lowell's early communication with newspapers and other general-interest publications: "It seems to be the first duty of those who are writing for such a public to be extremely cautious not to mislead. . . . Conjectures should be carefully separated from acquired facts; and the merely possible should not be confused with the probable, still less with the absolutely certain."[7]

Where "professional" astronomers refrained from participating in the debate on Mars' meanings, however, other scientists who conceived of themselves as public intellectuals like Lowell were quick to respond. Perhaps most eminent among them was the British naturalist Alfred Russel Wallace, who tangled with Lowell in a series of publications around the height of the Mars canal sensation in 1907. Wallace, who had jointly proposed natural selection as the mechanism of organic evolution with Darwin in 1858, took up numerous topics outside evolutionary biology over the subsequent half-century. In publications on themes ranging from land nationalization to vaccination, Wallace saw his intellectual project as interpreting the "social, political, and economic implications of evolutionary thought for the pressing questions of Victorian industrial society." Sharing the viewpoint with Lowell (and with Herbert Spencer) that it was a moral duty for scientists and intellectuals to "engage directly in social and political controversies," he published nearly 800 articles and books, lectured widely, and was a vigorous correspondent with scientists and political leaders alike.[8]

In the final years of his life, Wallace felt compelled to enter the debate over Mars, publishing a book-length rebuttal to Lowell in 1907 that provoked numerous responses and fanned the flames of Mars enthusiasm. Having published in 1903 a scientific justification for belief in Earth's uniqueness as the only inhabited planet, Wallace read Lowell's 1906 book *Mars and Its Canals* as a text that directly challenged his arguments and required a vigorous response. Inspired to write a longer critique than would be suitable for a journal or magazine, he asked his publisher (Macmillan, who had also published Lowell's 1906 volume) to publish it as a stand-alone book:

> I find that my article on "Mars" will be too long for a Review article, & I wish to have it published as a booklet. . . . It will be, as I mentioned before, what I consider a destructive criticism of Lowell's *theories*, while doing the fullest justice to his admirable work on the planet. The subject has attracted so much attention in almost every paper and magazine in the kingdom, that I think a *careful* and *thorough* yet *popular* exposition of the *facts*, and the *results* they lead to, will have a large sale at a low price. . . . I write therefore to know whether it would be against your rules or too unusual, for you to publish a criticism of one of your own books.[9]

After consulting with astronomer and physicist associates regarding some of his arguments, Wallace prepared a detailed critique of Lowell's hypothesis.[10] Following Lowell's own example, Wallace argued that the theory was best evaluated by nonastronomers, saying from the outset that only "a general acquaintance with modern science" was necessary to evaluate several of Lowell's claims. The argument regarding the artificiality of the canals, he suggested, "requires only care and judgment in drawing conclusions from admitted facts."[11] As he had hoped, Wallace's *Is Mars Habitable?* reached a wide readership and provoked intense discussion on the topic of Mars' habitability.

Some historians have praised Wallace as the voice of scientific reason, calmly and rationally dismantling Lowell's arguments about the temperature of Mars and delineating the limits of analogy's use as a methodology. Others have largely characterized Wallace as a theist with anthropocentric views that were dismissed by most of his scientific peers.[12] His perspective on evolution had famously diverged from Darwin's by exempting humans from the processes of natural selection, suggesting instead that the complexities of human intelligence might be due to some other process more akin to design. Wallace's opposition to Lowell's hypothesis, then, has been linked directly to his commitment to Earth's uniqueness and the theism this commitment implies. But *Is Mars Habitable?* is much more than a statement of theist belief or a treatise on atmospheric temperature influences. It makes a number of arguments about the nature of irrigated civilization, musters biological arguments about the conditions for life, and uses then-recent geological theory to offer a scientifically plausible natural explanation for Martian surface patterns. Closer examination of these arguments and their popular reception helps place the Lowell-Wallace debate over Martian canals within a richer intellectual context.

Physical Geography: A "Vast Sahara"

One of the few points on which Lowell and Wallace agreed was the arid condition of the red planet, with Lowell's "vast Sahara" echoed in Wallace's "terrible desert."[13] Although the two men proceeded from this point of agreement to widely divergent interpretations of Martian physical geography, the fundamental importance of this agreement should not be overlooked. During the era of heightened interest and belief in the Martian canals, discussions regarding the red planet's habitability and the extent of its analogy to Earth hinged on the fact that Mars was assumed to be a desiccated desert world characterized by the extreme conditions of its physical environment. These representations of Mars as arid were neither coincidental nor unimportant.

At the turn of the twentieth century, the narrative of Martian aridity was actually fairly new. Earlier speculative works had painted Mars as a lush, watery landscape, with its dark patches originally assumed to be water. Schiaparelli's groundbreaking 1878 canal map reflected this deeply entrenched assumption by using a light-blue coloring for the darker areas to indicate their watery nature (refer to fig. 2.8 above). In 1892 French astronomer Camille Flammarion's spectacularly successful book on the state of Mars-related knowledge, which had an immediate impact on the English-speaking world of astronomy, made numerous references to the assumed aqueous nature of the red planet.[14] By the mid-1890s, however, these portrayals had been irrevocably reversed. The oceans had been recast as tracts of sparse vegetation, the continents had been reformulated as "one vast desert waste," Pickering's circular "lakes" had been transformed into vegetative "oases," and the irregular water's-edge appearance from Schiaparelli's first map had given way to an increasingly geometric appearance.[15] (Refer, for example, to figs. 2.10 and 2.11 above.) Because the emergence of a desert narrative for Martian geography coincided with the growing popular sensation over the planet's habitability, the origins and nature of this geographical narrative merit further attention.

The switch to a desert narrative was driven primarily by the data and arguments of Percival Lowell. After he established his observatory in 1894, Lowell's first major contribution to the scientific understanding of Mars was an analysis of variation in the visibility of the lines/canals on its surface. Though many astronomers had previously reported that the canals did not seem to be equally visible at all times throughout the year, no one had conducted any systematic study or offered a comprehensive analysis of this phenomenon. Relying on Pickering's recent proposition that the dark areas on Mars could be vegetation rather than water, Lowell hypothesized that variations in canal visibility indicated seasonal changes in vegetative growth. To assess the patterns exhibited by such variation, Lowell made a detailed report of canal visibility throughout the months preceding and following the 1894 opposition, organizing his data by Martian latitude. In this geographic format, Lowell's data showed that the changing visibility of the canals followed interesting seasonal patterns: canals first began to appear near the Martian pole in Martian spring, then appeared to extend in length toward the Martian equator throughout the summer. In fall and winter, the reverse pattern occurred, as the canals receded in visibility from equator to pole.[16] (See figs. 5.1 and 5.2 for Lowell's graphical display of data related to this phenomenon.)

Assuming that the visibility of the canals indicated a seasonal growth of vegetation, Lowell pointed out that the Martian pattern was exactly opposite to that which prevails on Earth, where vegetative growth proceeds from equator to pole on the heels of seasonally warming weather. The inverse relationship of canal visibility to latitude, Lowell asserted, must rule out temperature as the limiting factor in vegetation growth on Mars. Instead, he argued, availability of moisture was the more likely limit, with his "cartouche" patterns confirming that Martian vegetation was watered solely by seasonal melting of the polar ice caps.[17] The polar caps' obvious waxing and waning was already one of the oldest and most firmly entrenched elements of the basic Earth–Mars analogy, indicating an active hydrologic circulation on Mars. In contributing his interpretation of canal visibility, however, Lowell radically asserted that the polar caps were the *only* source of water on Mars. His data seemed to show that water flowed from the poles toward the equator, meaning that the polar caps were key to understanding the geometric patterns girdling the planet. In Lowell's hypothesis, the geometric lines comprised an ingenious network of irrigation canals built by intelligent

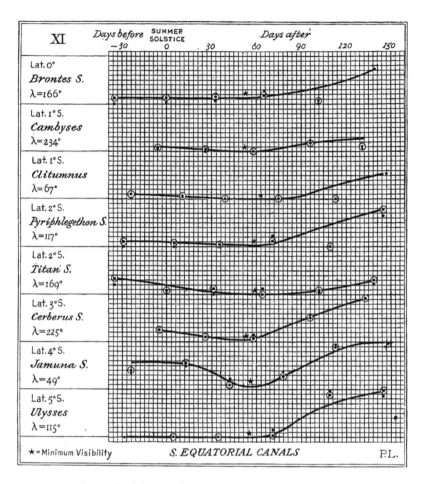

Figure 5.1. "Cartouches" showing changes in seasonal vegetative extent on Mars, 1904.
In this simple format, published in the *Bulletins of the Lowell Observatory* (Lowell,
"Cartouches," on p. 74), Lowell charted the changing visibility of the Martian canals
by latitude, showing that vegetative growth followed a pattern inverse
to that which would be expected if temperature were the limiting factor.

inhabitants to cope with the effects of extreme aridity. In his system, spring-
time snowmelt from the polar caps was conveyed by "gravity" and artifice to
the "tropic zones," where it watered a parched landscape, eventually evapo-
rated into suspended water vapor, and was then circulated by light air currents back toward the poles for wintertime deposition as ice.[18] The relatively
frail swaths of vegetation running alongside the canals were nearly lost in a
landscape Lowell described as "really one vast Sahara, a waterless waste."[19]

As Lowell himself acknowledged, this elaborate theory of Martian civilization was critically dependent on the idea that Mars was a desert world:

The discovery that there was no surface water on Mars was one of the fundamental factors in the . . . theory of Martian phenomena; for this fact proved Mars to be a desert world—five eights of its surface actual Sahara and the other three eighths but ill supplied with what is all essential to a world as we know it—moisture.[20]

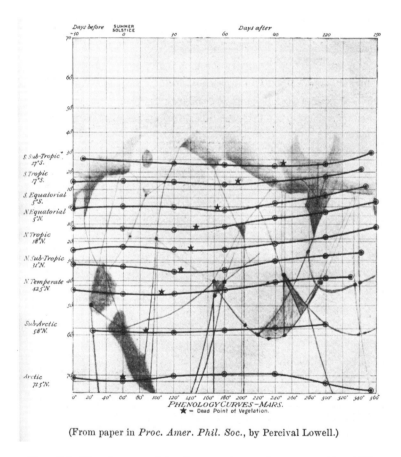

(From paper in *Proc. Amer. Phil. Soc.*, by Percival Lowell.)

Figure 5.2. "Phenology curves" showing vegetative growth patterns on Mars, 1906. Combining in one display the cartouches he had created for all Martian latitudes, Lowell assembled this graphic for *Mars and Its Canals* (on p. 343) to show that vegetative growth began near the poles during each Martian hemisphere's spring season.

In presenting his revised version of Martian physical geography, Lowell made effective use of terrestrial landscapes for comparison. The very visibility of the Martian surface was easily equated with the cloudless skies known to dominate Earth's desert regions. Although there were infrequent reports of a thin, veil-like haze on Mars from time to time, no one could dispute that virtually all of the Martian landforms were clearly visible at any given time. This was in stark contrast to the situation for Venus, where the planet's aspect changed so frequently that astronomers determined the entire planet was enshrouded by thick clouds. The pale colors observed on Mars, likewise, were assumed to find their closest analogy in the same desert landscapes:

> [T]he ochre regions . . . seem to be nothing but ground, or, in other words, deserts. Their color first points them out for such. The pale salmon hue, which best reproduces in drawings the general tint of their surface, is that which our own deserts wear. The Sahara has this look; still more it finds its counterpart in the far aspect of the Painted Desert of northern Arizona. To one standing on the summit of the San Francisco Peaks [near Flagstaff, Ariz.] and gazing off from that isolated height upon this other isolation of aridity, the resemblance of its lambent saffron to the telescopic tints of the Martian globe is strikingly impressive.[21]

Although Lowell often used the Arizona landscape around Flagstaff to establish his Mars-related desert analogies, the general discussion of Martian deserts and oases seems to have evoked a specifically North African or Middle Eastern landscape for many commentators. Mars' color was often described in comparison to the Sahara, and the image of a waning ancient civilization was frequently invoked. (See the not-so-subtle inclusion of Egypt's sphinx in the illustration for a speculative newspaper article about the red planet in fig. 5.3.)[22]

Even Lowell himself admitted a strong parallel between the Saharan landscape and his vision of Mars. Having traveled to North Africa several times, both before and during his career as an astronomer, he had occasion to observe the constraints and rituals of oasis life. In a letter he wrote privately to his sister from French North Africa in 1896, Lowell explicitly linked the desert landscape with Mars:

> And then Arabs everywhere in picturesque squalor and beautiful bronze skins which constitutes so large a part of their clothing. Flowers too sold by the same for love, one may say, for the money paid is next to nothing. Then cafes, also, innumerable in the open air, their little French tables and iron-work

Figure 5.3. Feature story in the *New York Herald*, December 30, 1906. Prominent inclusion of the Egyptian sphinx in this article's main illustration indicates that popular audiences accepted a strong connection between the supposed Martian civilization and a glorified ancient North African civilization. Courtesy Lowell Observatory Archives.

chairs setting most contentedly about. Do you know, it is a fancy if you will but I feel as if I were vouchsafed half-visions of the Martians in their perpetually sun-lit planet and oasis-like life.[23]

The romanticization of arid North Africa as "picturesque" and ancient in these conceptualizations of Mars points to the importance and complexity of the desert analogy. Beyond merely supporting the characterization of Martian physical geography as arid, the use of desert analogies served to introduce fundamental environmental assumptions and powerful cultural stereotypes into the general understanding of Mars. Many of these will be addressed in detail throughout this chapter. For now, it is sufficient to note that most critics of Lowell's theory, including Wallace, did not reject his fundamental desert analogy. Lowell's detractors endorsed the idea that Mars was severely arid, agreed that Mars was virtually cloud-free, and engaged unreservedly in comparisons of Mars to specific desert landscapes on Earth.

In opposing Lowell's theory without dismissing his critical desert analogy, critics thus limited themselves to arguments on other topics that paradoxically served only to reinforce popular enthusiasm for Lowell's arguments. There was considerable discussion in the astronomical journals and science magazines, for instance, about the temperature on Mars, with many critics arguing that Mars was far too cold to host life.[24] In another long-running area of critique, opponents of the inhabited-Mars theory insisted that the small mass of Mars could retain only a thin atmosphere, which would be far too rare to support any life forms. In the absence of conclusive evidence, however, the contestation of Lowell's assumptions about Martian temperature and air density actually contributed to his arguments about the evolution of Martian civilization. Lowell's logic linked Martian inhabitants to the physical environment in which they had developed, assuming that the greatest cultural advancement would occur in the harshest survivable environment. If the desert planet was not only parched but also frigid and short of air, Lowell's arguments regarding the evolution of an advanced Martian race were all the more convincing. Lowell himself thus cannily agreed with some of his critics in admitting that Mars was cold and that its atmosphere was thin, but he was vociferous in maintaining that it was nonetheless hospitable to life. Wallace's *Is Mars Habitable?* provided the most substantial challenge to this position, systematically dismantling Lowell's claims about Martian temperature and providing a detailed discussion of the biological conditions necessary for organic life to develop and evolve. Wallace claimed that Mars was unlikely to be warmer than Earth's moon, owing largely to its lack of either an insulating atmosphere or substantial water bodies. Relying heavily on geographical research—particularly that conducted in mountainous and arid regions—he argued that temperature swings and extremely low nighttime/wintertime temperatures would dominate in areas with minimal atmospheric pressure or water vapor content. As far as life on Mars, Wallace was certain that the Martian temperatures "must be far too low to support animal life."[25]

Immediately in response to Wallace's analysis of Mars as far too cold and far too airless to allow for the development of life forms, Lowell reworked his arguments and restated his comprehensive theory of Martian life in 1908's *Mars as the Abode of Life*, his last book on the subject of Mars. In making delicate arguments about the harshness of the Martian landscape (i.e., that it was harsh enough to serve as an evolutionary honing device but not so harsh as to kill off all life), Lowell produced his most geographical book, supporting his points with extensive analogies drawn from geographical texts and research conducted in remote and desert locations. He used the American naturalist Clinton Hart Merriam's 1889 research on the biogeography of Arizona's San Francisco Mountains to argue that animal life can survive extremely low temperatures as long as summer temperatures are sufficiently high to support reproductive activities. Lowell even reported on fieldwork he himself conducted in the San Francisco Mountains after reading of German geographer Alexander von Humboldt's finding that the plateau-flanked northern side of Asia's Himalayan mountain range produced considerably higher temperatures than those measured on the south side, where the peaks were exposed above the Gangetic Plain.[26] Borrowing a Humboldtian image from Scottish geologist Archibald Geikie's *Elementary Lessons in Physical Geography* (see fig. 5.4), Lowell argued that contiguous land masses like plateaus served to moderate temperature and to support the movement and survival of animal populations under extreme climatic conditions. He presented his own data from the San Francisco Peaks (following Merriam, see fig. 5.5) to show the general applicability of this idea, which he argued could be applied to Mars with its presumed extensive plateau landscape and lack of peaks. Lowell also used these same authors and data, among others, to argue that the "thinning of air proves no bar to a species," citing the case of meadowlarks' comfort in both the American Great Plains and Rocky Mountains as well as the case of thriving human populations in high-elevation Andean cities like Quito, Ecuador.[27]

Although Wallace's 1907 critique had carried significant weight with both scientific and popular audiences, Lowell's reworked arguments served to swing popular opinion back toward support of the inhabited-Mars hypothesis, even if it fell short of reviving the old idea that Mars enjoyed

From Geikie's "Elementary Lessons in Physical Geography." (The Macmillan Company.)

VERTICAL DISTRIBUTION OF CLIMATE ON MOUNTAINS, SHOWING HOW
LAND-MASSES RAISE THE TEMPERATURE

Figure 5.4. Illustration of altitude-dependent vegetative zones, 1908. Lowell used this illustration from a geography textbook in his 1908 book *Mars as the Abode of Life* (on p. 98) to show that topography played a major role in influencing temperature. He argued that Mars would have fewer zones of climatic extremes than Earth, given its relatively level land surface.

a balmy climate. (One 1895 commentator had memorably suggested that Martian polar explorers must enjoy pleasant travel conditions that were supposedly comparable "to a summer jaunt through Norway and Sweden.")[28] On Lowell's Mars, the conditions were harsh but livable, comparable to terrestrial extremes. Throughout all of these debates over Martian physical geography, the Lowellian construction of Mars as fundamentally arid remained virtually unchallenged. Although some commentators at the time questioned whether the Lowellian theory was based on his observations of aridity or whether the observations of aridity were problematically based on his theory of inhabitants, Lowell's successful representations of Mars as an inhabited planet were inextricably linked to the construction of an arid physical geography for Mars.

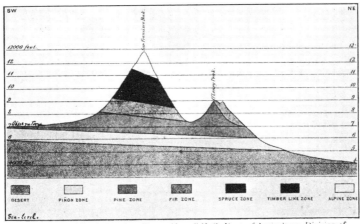

After a plate in " North American Fauna, No. 3," U. S. Dept. of Agriculture, Division of Ornithology and Mammalogy, by Dr. Merriam.

DIAGRAMMATIC PROFILE OF THE SAN FRANCISCO AND O'LEARY PEAKS, FROM SOUTHWEST TO NORTHEAST

The diagram shows the several life zones and the effects of slope exposure, but also shows what is unnoticed by the monograph, the effect of a plateau upon life. The location of the Lowell Observatory is indicated by the star.

Figure 5.5. Illustration of plateau effect on vegetative zones, 1908. Lowell used this figure in his 1908 book *Mars as the Abode of Life* (on p. 99) to illustrate his contention that temperatures decreased more slowly with altitude in proximity to a level plateau. Similar to fig. 5.4, the depiction of vegetation in altitude-dependent zones followed work by the well-known German geographer Alexander von Humboldt.

Landscape Change: The March of Martian "Desertism"

One of the more powerful stereotypes about deserts that emerged in Mars science was the idea that deserts were a degradation of forest landscapes, a sign that nature had been mistreated, and that they could be rehabilitated. In presenting Mars as a desert planet, Lowell fundamentally presented Mars as a dying planet, one that had lost its forests and oceans. This was a persistent idea, one that has stayed with us since Lowell's time—as applied both to Mars and to terrestrial deserts as well. The deep entrenchment and impressive longevity of the dying planet metaphor in Mars science and science fiction have been discussed thoroughly and capably by Markley.[29] In this section, I illustrate the ways in which this particular desert-related stereotype echoed the latest thinking in physical geography and also reflected the science associated with colonial administration and imperial expansion.

From his very first publications, Lowell postulated not only that Mars was arid, but also that it was actually undergoing an unrelenting process of *increasing* aridification. Lowell referred to the long-term climatic processes affecting Mars as "increasing terrestriality," "parching," or "desertism." He argued that the red planet had once hosted oceans as extensive as Earth's, but was well along an irreversible path toward becoming a "dead" world, like Earth's moon. In the intermediate stage, Lowell claimed, Mars had lost its oceans but still retained enough moisture to support limited vegetation. The planet's minimal water supply was said to be locked up in vapor form in the atmosphere and in solid form at the frozen polar caps, flowing over the surface as a liquid only infrequently. He declared that just as the Martian oceans had gradually vanished with time, Mars' land had begun to lose its water as well, rendering "once fertile fields" into devastated deserts. According to his observations, more than half of the Martian surface "is now an arid waste, unrelieved from sterility by surface moisture or covering of cloud. Bare itself, it is pitilessly held up to a brazen sun, unprotected by any shield of shade."[30]

Lowell's construction of inescapable Martian desertism relied heavily on Spencer's version of the nebular hypothesis, which was the leading model of cosmic evolution in the mid- to late nineteenth century. The nebular hypothesis held that all planets and heavenly bodies had formed from a common gaseous nebula. Upon formation, the individual planets were thought to begin an irreversible process of cooling and shedding moisture. The smaller the planet, the more quickly this evolutionary process was thought to occur, given the greater surface-area-to-volume ratio of small planets. Earth's tiny moon, for example, was known to be a completely dry and airless world, a perfect example of the end-state of planetary evolution. Mars, also on account of its small size, was likewise considered "an old world, a world well on in years, a world much older relatively than the earth," though it was not yet dead. Although the nebular hypothesis was somewhat past its prime by the time Lowell used it as the basis for his inhabited-Mars theory, he was thoroughly committed to the Spencerian model of predictable phases of physical evolution.[31]

The Spencerian model was also used extensively at that time in the emerging field of geomorphology, the study of landscape change. William Morris Davis, a geomorphologist who founded the Association of American Geographers in 1904, conceptualized landscape as the product of sequential erosion processes, drawing extensively on Spencer's application of concepts from organic evolution to other spheres of experience. Just as Davis considered both evolution and erosion of landscapes as "inevitable, continuous and irreversible process[es] of change producing an orderly sequence of transformations," Lowell fundamentally viewed Mars' condition as one of numerous stages of physical development.[32] He used anthropomorphic terms to describe the aging of Mars' landscapes in the same manner that Davis personified geology with references to its "life cycle."[33] In Lowell's terms, desiccation was akin to the inevitable decline and death of the body: "Standing as it does for the approach of age in planetary existence, it may be likened to the first gray hairs in man."[34]

To explain the sinister import of desertism for his earthbound audiences, Lowell's *Abode of Life* set forth numerous examples of aridification and desert growth on Earth. He cited James Dwight Dana's *Manual of Geology* for evidence that Earth's own oceans were steadily shrinking, including a map of North America that showed more than half the continent's landmass as having emerged from the ocean since "the close of Archaean time."[35] He also cited Ellsworth Huntington's work on climate change, in which the geographer argued that many of the Earth's driest regions had once been much wetter, showing a trajectory of desiccation on Earth.[36] In arguing that Mars' continents were subject to the same process, Lowell cited Huntington's own evidence of Earth's desiccation: "The Caspian is disappearing before our eyes, as the remains, some distance from its edge, of what once were ports mutely inform us. Even so is it with the Great Salt Lake, the very rate of its subsidence being known and measured."[37] In 1924 Huntington revised his hypothesis to acknowledge fluctuations or "pulses" in climatic history, thus abandoning the idea that the process of climate change was linear in the direction of desiccation. But during Lowell's time, a broad contemporary discourse regarding unidirectional landscape and climate change on Earth intersected perfectly with the story of an inhabited Mars.

Arizona's own Petrified Forest featured prominently in Lowell's arguments as one of his prized examples of terrestrial desertism. Having embarked on several fieldtrips to the Petrified Forest, Lowell reported that the scattering of petrified tree trunks among desert grasses and cacti provided silent witness to a historical age of plentiful moisture. Where giant trees had once stood in a dense forest, the tallest and densest plant life in northern Arizona was by then limited to the tops of mesas, such as that on which Flagstaff was perched: "Their lofty oasis is all that is now left of a once fertile country; the retreat of the trees up the slopes in consequence of a diminishing rainfall."[38] Lowell considered his field observations in the Petrified Forest to be quite important and even submitted an article for publication in the *Bulletin of the American Geographical Society*, an organization to which he had been inducted as a member in 1905.[39] Although this article was apparently never published, Lowell's primary research and findings on this topic had already been presented in *Abode of Life*.

Lowell's commitment to this view of Earth's deserts as degradations of prior forests spurred one of the most derisive public critiques he ever received. Eliot Blackwelder, a professor of geology at the University of Wisconsin, published a scathing opinion in *Science* in 1909, which accused Lowell of relying on outdated geological theories and ignoring recent work. Lowell's insistence that Earth's oceans were shrinking while its deserts were growing was an old idea, Blackwelder charged, further suggesting that "if Dana were alive to-day he would doubtless repudiate the idea, for it is wholly contrary to the mass of fact more recently made known." Furthermore, Lowell's observation-driven analysis of arid regions like the Petrified Forest produced "gross errors" in interpretation, Blackwelder argued, because it did not take into account recent findings that the geologic record was defined more by fluctuation than by linearity. Blackwelder characterized Lowell's argument as "astonishing and disastrous," concluding that "[c]ensure can hardly be too severe upon a man who so unscrupulously deceives the educated public, merely in order to gain a certain notoriety and a brief, but undeserved, credence for his pet theories."[40]

Even if Blackwelder and other geologists had begun to perceive a history of geologic and climatic fluctuation, however, Lowell's commitment to inevitable desertism as a foundation for his hypothesis stood him in good stead with nonspecialist audiences. Not only were generalists and popular audiences unlikely to be aware of the recent advances in geology, but they were much more likely to be aware of desert-related concerns in the colonial and imperial spheres, where geographers and other scientists were intensely engaged with the idea of desert expansion. As Diana Davis has shown in an important reinterpretation of French colonial science in North Africa, the trope of desertism (which eventually came to be termed "desertification") was an integral component of the colonial project. Throughout North Africa's Maghreb region, French colonial administrators depended on a scientific narrative of deforestation, aridification, and desert growth to wrest control of natural resources from local populations.[41] The budding science of forestry was particularly influential in lending credence to a narrative of desert growth that became so deeply ingrained that it continues to dominate today's understandings of the North African landscape, despite considerable advances in ecological science and ample evidence to the contrary.[42] Much of the news regarding colonial affairs, international trade, European expansion, and imperial competition thus presented an essentialized notion of deserts as degraded landscapes. For popular audiences, it would have been impossible to ignore this widespread trope, Blackwelder's disgust notwithstanding.

In comparing Martian landscape to terrestrial deserts, then, Lowell easily assumed the vocabulary and perspective that was widely used to characterize much of the Middle East. In fact, he directly applied the deforestation narrative as one of his primary analogies:

> Upon the southern coast of the Mediterranean, at the edges of the great Sahara, are to be seen to-day the ruins of vast aqueducts stalking silently across the plains. . . . At the present day the streams are incompetent to supply the aqueducts, the very presence of which attests that in the past this was not so. The land has parched since times so recent as to be historic, recorded by the monuments of man. . . . In a startling manner it brings before us the speed with which the desert is gaining on the habitable earth.[43]

> In the same manner streams descend from the cedar-clad range of the Lebanon to lose themselves in the Arabian desert just without the doors of Damascus; and Palestine has desiccated within historic times. Palestine, a land once flowing with milk and honey, can hardly flow poor water now, and furnishes another straw to mark the ebbing of the water supply.[44]

In Lowell's narrative, the desertism of Mars was presented as an ominous process leading to the production of "a world-wide desert where fertile spots are the exception, not the rule, and where water everywhere is scarce."[45] This was exactly the scenario widely assumed to have taken place in North Africa and its Saharan landscapes, where the ancient Roman ruins provided architectural testament to a land once fertile and forested.

In a departure from the colonial discourse, however, Lowell did not fault Mars' inhabitants for their planet's creeping desertism. Unlike the narratives that described much of the Middle East and Africa, in which local inhabitants were blamed for neglecting their landscapes and squandering the ancient paradise, Lowell explicitly absolved the Martians of any responsibility: "This making of deserts is not a sporadic, accidental, or local matter, although local causes have abetted or hindered it. On the contrary, it is an inevitable result of planetary evolution."[46] Lowell's focus on the role of natural forces in desert growth, rather than on the agency of desert inhabitants, perhaps reflects the coloring of his perspective by the American expansionist experience rather than the European imperial experience (discussed in more detail in the next chapter). Regardless of where he placed the blame, however, Lowell's startling picture of accelerated landscape change was powerfully influential in supporting his theory of Martian inhabitants. By imagining Mars as a planet marked by its growing deserts, Lowell set the stage for his sensational inhabited-Mars theory: "To let one's thoughts dwell on these Martian Saharas is gradually to enter into the spirit of the spot, and so to gain comprehension of what the essence of Mars consists."[47]

Wallace did not directly address Lowell's construction of increasing aridity on Mars, but his arguments in *Is Mars Habitable?* accepted some of the fundamental assumptions on which the construction relied. Having argued that the biological preconditions for organic life were not met on cold and airless Mars, Wallace finished his book with a counterexplanation of Martian surface patterns that was meant to challenge Lowell's theory of intelligent civilization. In Wallace's view, the lines on Mars—which he assumed to be real on the basis of their frequent sighting by numerous astronomers—must have had some natural origin. He found an explanation for this origin in the recently mooted Chamberlin-Moulton planetesimal hypothesis of solar system formation, which focused on explanations of planetary history that were consistent with recent geological research on Earth.[48] Where the nebular hypothesis held that planets had formed through the condensation and cooling of hot, gaseous nebulae, the planetesimal hypothesis postulated instead that planets had formed by the continuous collision and accretion of small particles during periodic passages through fields of asteroids.

To Wallace (and others), the most significant difference between these two theories was in the amount of heat each supposed possible in the early stages of planetary formation. Where Laplace's early solar system was molten, Chamberlin's was cool, although the planetesimal hypothesis allowed that the meteoritic bombardments responsible for augmenting planetary mass during each orbital conjunction would have temporarily heated and liquefied planetary surfaces. Wallace argued that this periodic effect would not have produced enough warmth to incubate organic life on Mars through its long requisite development period, although it would have caused massive fracturing of the Martian landscape as molten surface material began to cool and shrink against a still-cool planetary core. In proposing this alternative scenario, Wallace accepted the basic Lowellian construction of Martian surface cooling that was so fundamental to the idea of increasing aridity. In this, Wallace was like many other critics who failed to challenge one of Lowell's fundamental claims and accepted the basic idea that all planets were losing heat. "Because their alternatives to the canal thesis assumed some version of [the same] evolutionary cosmology, Lowell's critics were forced to counter his theories using the same assumptions on which his case rested."[49]

Even in their reliance on different theories of planetary evolution, Wallace's and Lowell's arguments both reinforced the idea of an arid, dying planet. Lowell's construction of inhabitants that were involved in a life-and-death struggle on their dying world, however, clearly prompted the greater interest and sympathy from wide popular audiences. Lowell's "sublime, even tragic, depiction" of a planet's evolutionary transition to complete terrestrialism presented "aesthetic and philosophical implications" far more compelling than Wallace's arguments about seemingly mundane geological fractures on a dead world.[50] The dying planet construction echoed the latest thinking in physical geography and also reflected the science associated with colonial administration and imperial expansion. This very specific context shaped the nature of the "place" Mars had become and the way that audiences responded to it. On Lowell's Mars, the desert condition was not a temporary or local phenomenon. It was the product of an inalterable process of planetary decay that applied to Earth just as surely as it applied to Mars. "To the bodily eye, the aspect of the disk is lovely beyond compare; but to the mind's eye, its import is horrible. . . . For the cosmic circumstance about them which is most terrible is not that deserts are, but that deserts have begun to be. . . . They mark the beginning of the end."[51]

Cultural Geography: Social Darwinism and Martian Determinism

In proceeding from constructions of an increasingly arid physical geography to a narrative of advanced Martian civilization, Lowell relied on existing enthusiasm for an Earth–Mars analogy and on widespread acceptance of a landscape-society link. Noting that the chance of multiple perfectly straight watercourses intersecting at a perfectly circular lake was "millions to one" in nature, Lowell argued that the canals appeared "supernaturally regular" in straightness, width, and "systematic radiation from special points." Relying on an understanding of processes known to affect Earth's geography, Lowell stated that "[p]hysical processes never, so far as we know, produce perfectly regular results. . . . Too great regularity is in itself the most suspicious of circumstances that some finite intelligence has been at work."[52] Using basic analogical reasoning, then, Lowell insisted that the incidence of straight lines and circular intersections on Mars indicated certain evidence of intelligent beings:

> The whole system is trigonometric to a degree. If Dame Nature be at the bottom of it all she shows on Mars a genius for civil engineering quite foreign

to the disregard for prosaic economy with which she is content to work on our own work-a-day world. Her love for elementary mathematics is evidently greater than is commonly supposed.[53]

The design has all the geometric precision of one of those delicately constructed spider-webs which become the more admirable the more minutely they are scanned. As one's vision grows steadier, this mesh of linear markings takes on a more and more artificial look.[54]

Lowell was supported in this argument by the zoologist Edward Morse, an old friend who had traveled with him in Japan and who wanted to repay Lowell for his earlier support of Morse's own "crusade to preserve evidence of traditional [Japanese] culture in the face of rapid modernization."[55] In defending the theories of his ally, Morse argued that the patterns observed on Mars' surface were much more geometric than any natural features observed on Earth. His 1907 book on Mars included diagrams of street and rail networks, and both photographs and sketches of natural crack patterns, contrasting the canals' resemblance to the manmade landforms with their dissimilarity to the less regular patterning of cracked mud, cracked asphalt, or tectonic faults (see figs. 5.6 and 5.7). Such intuitive geographical analogy obviously appealed to popular audiences, as Morse's and Lowell's comparative statements and imagery were repeated regularly in newspapers and general-interest books and magazines. See, for example, a feature story in *The World Magazine* on Morse's trip to Flagstaff to observe with Lowell (fig. 5.8).

Both Lowell and Morse used the landscape patterns of the American West to drive home their points about artificiality. Viewed from Mars, Lowell argued, any visible evidence of human activity on Earth would be limited to "such semi-artificialities as the great grain-fields of the West when their geometric patches turned with the changing seasons from ochre to green, and then from green to gold. By his crops we should know him,—a telltale fact of importance because probably the more so on Mars."[56] Morse likewise invoked analogy with the western landscape for comparison.

If in the mind's eye we were to survey the Earth from Mars the only feature we should find at all paralleling the lines in Mars would be found in the level regions of the West, where, for thousands of miles, the land extends in vast stretches. In these regions would be found lines of railroads running in straight courses, starting from definite places, converging to common cen-

PLATE V

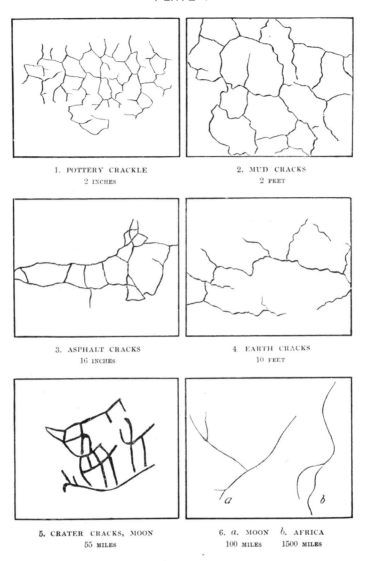

1. POTTERY CRACKLE
2 INCHES

2. MUD CRACKS
2 FEET

3. ASPHALT CRACKS
16 INCHES

4 EARTH CRACKS
10 FEET

5. CRATER CRACKS, MOON
55 MILES

6. *a.* MOON   *b.* AFRICA
100 MILES   1500 MILES

NATURAL LINES

CRACKS, FISSURES, ETC.

Figure 5.6. Illustration of natural crack patterns, 1907. Edward B. Morse used this illustration in his 1907 book *Mars and Its Mystery* (facing p. 112) to argue that natural cracking patterns showed none of the regularity observed in the linear features on Mars.

PLATE VI

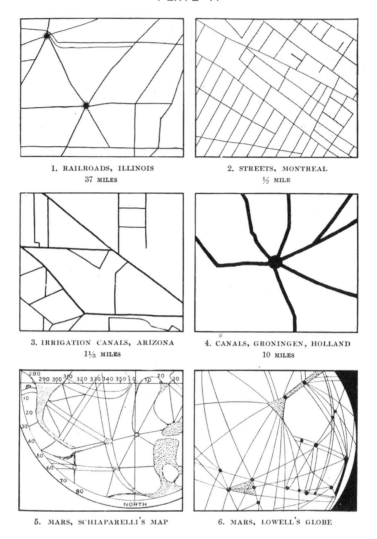

1. RAILROADS, ILLINOIS
37 MILES

2. STREETS, MONTREAL
½ MILE

3. IRRIGATION CANALS, ARIZONA
1⅛ MILES

4. CANALS, GRONINGEN, HOLLAND
10 MILES

5. MARS, SCHIAPARELLI'S MAP

6. MARS, LOWELL'S GLOBE

ARTIFICIAL LINES

RAILWAYS, STREETS, CANALS, ETC.

Figure 5.7. Illustration of manmade transportation networks, 1907. Edward B. Morse used this illustration in 1907 his book *Mars and Its Mystery* (facing p. 113) to show that artificial or manmade features had much more in common with the observed features on Mars than any naturally occurring crack patterns. (Compare to fig. 5.6, which immediately preceded this plate in Morse's book.)

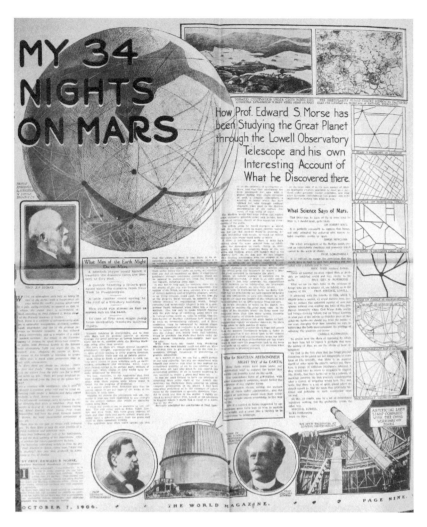

Figure 5.8. Feature story on Morse's visit to Lowell Observatory, 1906. This story, which appeared in the October 7, 1906, number of *The World Magazine*, sensationally reported on Morse's visit to Flagstaff, where he observed Mars for more than a month at the Lowell Observatory. Many of the photographs and schematic illustrations of geometric landscape patterns included alongside this story also appeared in Morse's 1907 *Mars and Its Mystery*. Courtesy Lowell Observatory Archives.

tres, their sides, in certain seasons, conspicuous with ripening grain fields, or again the work of the United States Reclamation Bureau running its irrigation canals in various directions throughout that great region. Both these kinds of lines would be artificial and both designed for purposes of conveyance—in the one case, merchandise and passengers, in the other case, water.[57]

As one who had never before observed Mars or published on the topic of planetary astronomy, Morse was not given any serious consideration by professional astronomers. In the popular press, however, his reception was fairly warm, indicating a broad acceptance of the artificiality argument and of Lowell's fundamental rethinking of the Earth–Mars analogy in terms of a class of features that could *not* be explained by terrestrial environmental processes.[58] A review of Morse's book in *The Nation*, for instance, credited Morse with demonstrating the importance Mars held for natural scientists and for showing that the planet was not inscrutable to lay observers:

> At first blush it might seem rather venturesome for a naturalist so eminent as the director of the Peabody Museum at Salem to appear before the world as an astronomical observer. But the field that Professor Morse has here entered, though belonging to the heavens, is not an essentially astronomical one in the sense of involving subjects or methods or observation with which a naturalist is not familiar.[59]

For those who followed Lowell's logic—that landforms unexplainable via known natural forces proved the existence of intelligent beings—the similarities between features on Mars and manmade features on Earth were overwhelming. Where the bizarrely geometrical map of Schiaparelli had first defied terrestrial analogy, Lowell transformed it into a familiar irrigated landscape, governed by known processes and technologies.

Not only did Lowell argue that the apparent artificiality of the Martian landscape proved the *existence* of a civilization, but he also asserted that the nature of that artificiality indicated the *level* of the civilization. In his 1906 book, *Mars and its Canals*, for instance, Lowell included a brief chronicle of "man's history," which highlighted landscape activity as the best indicator of evolutionary progress:

> While [man] still remained of savage simplicity, a mere child of nature, he might come and go unmarked by an outsider, but so soon as he started in to possess the earth his handicraft would reveal him. . . . It began with agriculture. Deforestation with its subsequent quartering of crops signalized his

acquisition of real estate. His impress at first was sporadic and irregular, and in so far followed that of nature itself; but as it advanced it took on a method-ism of plan. . . . Regularity rules to-day, to the lament of art. The railroad is straighter than the turnpike, as that is straighter than the trail. Communica-tion is now too urgent in its demands to know anything but law and take other than the shortest path to its destination. Tillage has undergone a like rectification. To one used to the patchwork quilting of the crops in older lands the methodological rectangles of the farms of the Great West are painfully exact. Yet it is more than probable that these material manifestations would be the first signs of intelligence to one considering the earth from far.[60]

The enthusiasm with which popular audiences and intellectuals in various fields engaged with this idea indicates its critical importance to the over-all sensation over Mars. It also suggests broad acceptance of the assumed landscape-society link, an idea then at the forefront of the emerging science of geography.

Beginning in the second half of the nineteenth century, geography had begun to gain a scholarly identity in Europe, where several university chairs in geography were established in 1870s Germany. Where geography had previously been considered primarily a practical tool, its increasingly im-portant role in the facilitation of German (and wider European) imperial expansion had propelled this elevation to "scientific" status.[61] Such baldly practical origins, however, left the new discipline with something of an infe-riority complex within the academy, particularly as the process of "academic atomization" in the European universities promised little structural support for such an integrative science. Within this institutional context, the new geographers needed to present their science as a search for universal geo-graphical laws. Toward this end, the influential German geographer Friedrich Ratzel suggested that evolutionary theory could provide the foundation for such laws and their application to studies of human–environment relation-ships. Relying heavily on Spencer, Ratzel conceptualized environmental conditions as the primary determinant of human progress through natural evolutionary "stages" of civilization.[62] In Ratzel's environmental explana-tion of human history, human evolution and cultural progress proceeded at different rates and in different forms at different places on the Earth's surface. As articulated in his *History of Mankind*, the "less favourable condi-tions" of the temperate zones had led to greater evolutionary progress and higher forms of civilization because there "[man] had to look after himself with more care than in the soft cradle of the tropics."[63] Many in the growing generation of geographers that followed Ratzel, particularly Ellsworth

Huntington and Ellen Churchill Semple in the United States, accordingly attempted to "read history through environmental spectacles," arguing that climate and landscape morphology were the primary determinants of cultural development.[64] Together, these geographers helped set the agenda for their discipline and defined its subject of study as the impact of the natural environment on human history and culture.

This linking of landscape and society not only provided an academic foundation for the discipline of geography, but it also gained sympathy from popular audiences influenced by the geopolitical contexts of Western imperialism. As Livingstone has argued, the theory of environmental determinism was in no way a departure from geography's practical roots as a tool of empire. The characterization of tropical peoples as helplessly backward, inferior, or immoral on account of their environmental situation conveniently served to provide a scientific mandate for imperialism, justifying even violent European intrusion into tropical realms as a commendable extension of higher civilization to areas of cultural deprivation:

> [D]iscussions of climatic matters by geographers throughout the nineteenth century and well into the twentieth century were profoundly implicated in the imperial drama and were frequently cast in the diagnostic language of ethnic judgement. To put it another way, the idioms of political and moralistic evaluation were simply part and parcel of the grammar of climatology.[65]

In theorizing the impact of natural environment—particularly climate—on native peoples, European geographers tended to focus on the tropical environments and peoples in which and with whom their governments and militaries were engaged: Africa, South Asia, Southeast Asia, the Caribbean, and Latin America.[66] American geographers, on the other hand, tended to focus more on the question of how westward expansion of European peoples across a temperate but increasingly arid continent would impact those who had presumably already achieved the highest levels of civilization. According to Semple, the American encounter with western landscapes had "modified European institutions and character," producing "the first genuine Americans." American historian Frederick Jackson Turner's infamous 1893 "frontier thesis" had likewise held that American society and character were fully defined by the encounter with frontier landscapes.[67]

Environmental determinism—as legitimized by academic geographers who positioned themselves within wider debates about the general applicability of evolutionary theory—was thus a critical part of the imperialist and expansionist narratives that proliferated in the Western world. And unlike new theories of stellar composition or geologic cycles, environmental determinism could be understood easily by wide popular audiences that fully supported its practical applications in the service of empire. When Lowell presented his view of Martian culture, it was not much of a stretch for his audiences to accept the basic principle on which it rested: a deterministic link between landscape and society.

Before ever publishing a single word about Mars, in fact, Percival Lowell was already well known with American audiences for his environmentally deterministic views on Asian culture. He had spent almost ten years traveling to Japan and Korea, had served as a diplomatic appointee in Korea, had gained some notoriety on the lecture circuit, and had published four successful books and numerous articles that explained Asian culture as paradoxically advanced yet inferior because of the specific circumstances of Eastern geography. In his best-known book on Japan, for instance, Lowell characterized the island nation's inhabitants as a product of their landscape:

> The torpor of the East, like some paralyzing poison, stole into their souls, and they fell into a drowsy slumber only to dream in the land they had formerly wrested from its possessors. Their birthright passed with their cousins into the West. . . . Artistic attractive people that they are, their civilization is like their own tree flowers, beautiful blossoms destined never to bear fruit.[68]

The underlying views that supported this interpretation also defined Lowell's approach to Mars, in which the impressive Martian civilization was defined by its environmental encounter. In an excellent biography, Strauss argues that Lowell's publications on both Asia and Mars were part and parcel of the same intellectual project: an expansion of Spencer's theory of evolution and the unity of the cosmos, as elaborated by Huxley, Fiske, and Haeckel.[69] Environmental determinism was merely the most effective route toward this philosophical ambition.

Although Lowell's role as both scientist and interpreter of such exotic lands as Japan and Mars has been commented at length, it is less well known that he further elaborated his deterministic beliefs in arguments about landscape-society links in his home country. In numerous politically themed lectures delivered after his astronomical career had begun, Lowell characterized American progress as no less environmentally determined than Japanese cultural stagnation. In a 1901 Independence Day speech in Flagstaff, for example, he linked American success to the historical encounter with North American landscapes:

That upon which we most pride ourselves, our shrewdness and our inventiveness is thoroughly climatic. We are quick because our nerves are tense through forces outside of ourselves. We are kept keyed up to our capacities, if not beyond them. Endeavor is of the very breath of our nostrils. There never, indeed, was a clearer case of adaptation to new conditions.[70]

The converse implication of this praise for Americans not only echoed the European geographers' construction of tropical peoples as sluggish and incapable of higher social organization, but it also asserted that Old World cultures were prevented by their long geographical tenure in Europe from achieving the advances that were possible in new American environments. Lowell thus characterized America's climatic debt exactly as did American geographer Semple: "In the large, fresh environment of the American continent the English race had been born again and now was animated with the irrepressible vigor of a youthful people."[71]

For geographers like Semple, the tenets of environmental determinism contributed directly to a politically powerful Social Darwinist vision of racial hierarchy as natural law. Social Darwinism, the interdisciplinary attempt to apply evolutionary theory to humans, conceptualized human societies as competing against one another in a struggle for survival that produced a Darwinian selection of the "fittest" societies. As Social Darwinism became a powerful explanation for Western imperial success, extending deep into both the intellectual and political spheres of Europe and America, geographers provided a "naturalistic explanation of which societies were fittest in the imperial struggle for world domination." Politicians enacted Social Darwinism during this same time period in several ways. Most notoriously, Social Darwinism led to attempts to control human evolution or even eliminate "unfit" groups through eugenics. Most pervasively in the United States, it supported a Progressive political movement that favored a class of technocratic elites invested with the authority and responsibility for controlling the environments (both natural and urban) that shaped individual Americans.[72]

Lowell's political speeches reflected his fundamental commitment to a Social Darwinist view of cultural hierarchy and to its expression in a specifically Progressive politics in the mold of Roosevelt. He vigorously opposed immigration, the unionization of American labor, and the advance of socialism on the grounds that they would delay the natural evolution of a superior American "race," increase violence, suppress individualism, and reduce American society to its lowest common denominator.[73] His objection to immigration and his support for eugenics echo the neo-Lamarckian work of American geographers, extending an argument he had made throughout his careers as both orientalist and astronomer. For Japan, Lowell attempted to characterize Asian development levels on a hierarchical scale "from savage to civilized."[74] For Mars, he used the intelligent canal-digging engineer as an exemplary confirmation that those societies at the upper end of the natural racial hierarchy were destined to come out on top of evolutionary competitions hastened by environmental crisis. His representations of Mars thus served to validate the Progressive political view that was presented very clearly in his unpublished political lectures. In order to best serve the interests of the greater society, Lowell argued, the naturally best and brightest of its citizens must be allowed to maximize their individual talents and wealth.

The political significance of Lowell's Mars work perhaps clarifies what many historians have seen as his irrational devotion to speculative claims in defiance of the accepted professional standards of scientific writing. It also justifies his use of moralistic and prescriptive writing styles—the same voice he had adopted for his books about Japanese and Korean culture. If we re-evaluate Lowell's "message" about inhabited Mars in this context, we can better understand the reasons it appealed to widespread audiences, despite public skepticism from leading astronomers. Because of its deterministic and evolutionary significance, the Lowellian story about Martian geography could be conceptually integrated into ongoing discussions about the nature of racial difference, the difficulties of cultural contact, the justifications for imperialism, and the role of science and technology in guiding continued Western expansion. Re-examination of the historical record shows, in fact, that it was these very geographical concerns that featured most prominently in popular representations of Mars and prompted lengthy comment from leading intellectuals outside the discipline of astronomy. Prominent American sociologist Lester Frank Ward, for instance, saw great value in the "lessons" to be learned through careful consideration of the Martian evolution scenario so integral to Lowell's hypothesis. In an enthusiastic reaction to Lowell's inhabited hypothesis, Ward accepted the idea of evolution through environmental constraints and viewed the increasing aridification of Mars as a powerful driver of its civilization's advance. The development of a Martian "race of beings of great industry and high intelligence," in Ward's view, foretold an optimistic future for the younger and less evolved Earth, despite the magnitude of the Martian water crisis: "the contrast with that old decadent orb that is now telling us its story, instead of depressing us, should inspire us with thankfulness that we are young, with faith in an unlimited future, and with buoyant aspirations for the progress of humanity."[75]

Wallace, by contrast, found Lowell's inhabited-Mars theory to be sociologically and politically illogical in its assumptions about human–environment linkages. Although the majority of Wallace's book *Is Mars Habitable?* focused on the physical conditions of the red planet's habitability, he also made a few pointed remarks about the nature of the civilization Lowell had posited on its surface. He rejected Lowell's idea that resource scarcity would lead to social harmony, for instance, identifying regional isolation as a much more plausible scenario for a world with limited and dwindling water supplies: "How, with such a desert as he describes three-fourths of Mars to be, did the inhabitants ever get to know anything of the equatorial regions and its needs, so as to start right away to supply those needs?" Similarly, Wallace pointed out that the sophisticated technological control of energy supposedly needed to build the canals would require a dense population "with surplus food and leisure enabling them to rise from the low condition of savages to one of civilization, and ultimately to scientific knowledge."[76] In these and other comments in the book's brief conclusion, Wallace revealed an important (and nontheological) opposition to Lowell's assumptions about the landscape-society link.

This opposition can be read partly as an expression of Wallace's mature beliefs about the nature of evolution and its effects on humans. Many of his contemporaries and subsequent commentators have criticized Wallace for characterizing humans as a special case, usually attributing his belief in human and terrestrial uniqueness to theological blinkerism.[77] One of Wallace's most recent biographers clarifies, however, that Wallace did not see the human species as exempt from evolution so much as he believed that modern society prevented the continued operation of natural selection. Wallace's writings asserted that evolutionary processes *had* operated on ancient cultures, serving to produce greater capabilities, better intelligence, and higher forms of civilization over time. With the advent of capitalism and the reorganization of Western societies in the industrial age, however, Wallace felt that natural selection had necessarily ceased its operation, causing stagnation in civilization's advance. In Wallace's view, the capitalist concentration of wealth was particularly disruptive of the evolutionary process, ensuring that some people could easily produce progeny regardless of any mental or physical merits. The modern economic system and the Victorian culture it spawned thus "frustrated, rather than facilitated, genuine evolutionary advance." This dilemma could not be resolved through the eugenics proposals then in vogue and under discussion, Wallace argued, given that such practice would be "socially ineffective and evolutionarily insignificant."[78]

Partly because of this intellectual position and because of his own observations of increasing "misery" in industrial England and America, Wallace was extremely sympathetic to the socialist critique of capitalism. Upon returning from years of research in the Malay Archipelago to the United Kingdom, Wallace's confidence in European racial superiority began to waver. Confronted with highly industrialized urban landscapes, Wallace grew concerned that "technical mastery over the forces of nature" had produced great wealth but also an explosion of poverty, crime, and inequality. In his view, it was the "Europeans, rather than the so-called savages among whom he had lived, [who] suffered under a 'barbaric' social and moral organization." As a result, Wallace became a committed socialist and activist for land nationalization, writing widely on sociopolitical topics while he also continued his scientific work on evolution. Fichman has identified Wallace's larger intellectual project as "biological socialism," which sought a "moral and cultural regeneration of Victorian society" and a "remedy for . . . the deplorable plight of the working class."[79]

In an intellectual move that was clearly influenced by these political beliefs, Wallace reversed his original opposition to Darwin's postulation of sexual selection as an agent of evolution. Theorizing that modern capitalism was the culprit that had produced a stagnant society, Wallace argued that only socialism could restore humanity's advance. In a more egalitarian society, he postulated, female sexual selection would become the agent of continued human evolution and improvement, as it would be based on physical or mental superiority, rather than wealth:

> Socialism, by removing disparities of wealth and rank, would eliminate the economic and political prejudices that, Wallace claimed, dominated the selection of reproductive partners in Victorian society. In their place, mate choice would focus on those higher moral and intellectual traits often neglected (or rendered subservient) in competitive capitalist society.[80]

This was a fundamental change in Wallace's thinking, signaling his "convergence of evolutionary biology and sociopolitical reformism."[81]

This stage of Wallace's own intellectual evolution coincided with his entry into the debate over Mars, and its influence on his opposition to Lowell should not be overlooked. Wallace's socialist vision of an egalitarian society as the only route to higher evolution stood in stark contrast to Lowell's Progressive vision of a technocratic elite emerging from an environmentally deterministic evolutionary process on Mars. Where Wallace rejected Lamarckian explanations of evolution, Lowell postulated a world fundamentally defined by environmental conditions and influences. Where Wallace was an advocate for socialism through land nationalization, Lowell offered Mars as an example of how global peace could be achieved through massive technological projects controlled by a "benevolent oligarchy of the intellectual elite."[82] These differing and intertwined views on politics and evolution help explain the gulf of disagreement between Wallace and Lowell over Mars.

Most commentators on Wallace's role in the Mars debates have focused on his arguments regarding the physical conditions for planetary habitability and the logical acrobatics he performed to assert that Earth was likely to be the only "habitable" planet in the universe. Markley, particularly, accuses Wallace of hypocritically ignoring his own logic of natural selection and making somewhat tortured statistical statements such as:

> If the physical or cosmical improbabilities as set forth in the body of this volume are somewhere about a million to one, then the evolutionary improbabilities now urged cannot be considered to be less than perhaps a hundred millions to one; and the total chances against the evolution of man, or an equivalent moral and intellectual being, in any other planet, through the known laws of evolution, will be represented by a hundred millions of millions to one.[83]

Wallace's concluding arguments in *Is Mars Habitable?*, however, are revealing for their directness in taking up Lowell's assertions about the theoretical evolution of Martian beings. When Wallace directly rejected Lowell's super-advanced Martian civilization, he did so not on the basis of any explicit theological argument but on the basis of a disagreement as to how human societies develop and how humans evolve under different economic systems. His challenge to Lowell's environmentally deterministic suggestion that resource scarcity would lead to accelerated evolution was based on a concern that technocratic capitalistic societies created conditions that actually prevented evolution from occurring. Wallace's exemption of capitalistic civilizations from natural selection was thus *not* a logical leap but rather a well-reasoned application of evolutionary theory to modern society. The political and philosophical perspectives of the two men thus found little common ground, exacerbating their sharp dispute over Mars.

Human–Environment Interaction: Irrigated Mars

Lowell's and Wallace's fundamental disagreement on the landscape–society link found specific expression in their comments about the existence, nature, and logic of an irrigation network on Mars. As should be clear by now, the postulation of irrigation canals on Mars was the single most sensational claim Lowell made, captivating public attention, driving media coverage, and provoking heated disagreement among scientists and commentators. But why, we must ask, would irrigation have emerged as a plausible interpretation of Martian markings in the first place, and how did the idea of Martian irrigation achieve such resonance among so many diverse audiences? Again, a focus on emerging trends in the discipline of geography helps provide critical context for these developments in Mars science.

For those early geographers who struggled to define their discipline as a study of human–environment relations, irrigation featured prominently in two primary intellectual strands. On the one hand, the presence of manmade networks for water control could be analyzed as evidence of environmental constraints on human action and on cultural development, with the complexity and efficiency of irrigation technology typically used as a visual proxy for cultural level. In this regard, irrigation was important to those environmental determinists who saw the human–environment equation as being weighted heavily toward environmental power. Lowell's prediction that Earth was "going the way of Mars" and would eventually "roll a parched orb through space," for example, drew from and amplified a certain terror regarding aridification.[84] American geographer Huntington was developing his deterministic understanding of the link between climate and culture at this same time, suggesting that several of Earth's great ancient civilizations had fallen when their climates began to shift toward aridity. Mars seemed to provide the perfect example of Earth's own future on this path, where "a race of vast antiquity and supreme wisdom [clung] desperately to the orb that bore it, half gasping for breath and hoarding every drop of its precious water, but [was] doomed in the relatively near future to face the lingering death of a dying world."[85] If Mars was any example, Earth's own increasing desert extent would bring continued challenges to civilization and to conventional understandings of human–environment interaction.

On the other hand, however, the sheer magnitude of large-scale irrigation and its fundamental alteration of arid landscapes supported an opposite theoretical position that considered humans to be much more powerful in the human–environment relationship. George Perkins Marsh's 1864 catalog of humanity's destructive impacts on the Earth (which inspired the first wave of conservation activism) highlighted especially the ways that human technology could effect massive hydrological change.[86] Although Marsh's voice was largely drowned out in the academic realm by the superior appeal of environmental determinism during the colonialist heyday, his perspective remained important in popular circles and soon re-emerged as a primary driver of research in geography and other disciplines. In fact, it was through a major conference and monograph dedicated to Marsh that geographers finally and officially put environmental determinism to rest in the mid-twentieth century.[87] Not surprisingly, this monograph used irrigation as one of its most prominent examples of man's deleterious effect on nature.[88]

In American geography and policy, questions regarding the role of irrigation in human–environment relations were especially important in the nineteenth century, given the young nation's celebrated conquest of a vast, arid territory to the west of its original territorial boundaries. At the same time that Huntington and Semple were writing about the impact of nature on humans and of the frontier's defining influence on the American national psyche, the early American conservation movement was born to protect frontier landscapes from man's growing impacts.[89] Thus, both intellectual strands of the early human–environment geography were in operation at the crucial "closing" of the American frontier in the late nineteenth century.

Faced with vast unwatered lands, westward-moving Americans and their institutions quickly prioritized irrigation as a necessary environmental mediation to allow settlement. But there was a lively debate as to how irrigation could best be implemented, given the political (and geopolitical) climate of the times. Worster identifies a widespread American "confidence that conquering the desert through irrigation would produce a more perfect democracy," as technological water control could open more land to more Americans, allowing everyone to contribute to the growing capitalist economy. Marsh, however, was one of few dissenters who saw irrigation as not only environmentally problematic but also unfair—a "potential threat to democratic ideals"—given the more likely scenario that an elite few would gain control over limited water sources. In the end, the rational-scientific aspects of irrigation through technology won public support for water projects that seemed to promise "progress toward a more rational society."[90] Toward the turn of the twentieth century, engineers—rather than frontier settlers or irrigation capitalists—became the most influential Westerners, despite the fact that few of those who turned attention to irrigation were actually from the West. As these engineers gained status within the American irrigation movement, they advocated for greater levels of centralization and government control to promote water-development schemes comparable to those reportedly operating in British colonial India and Egypt. In these arid realms, the British colonial governments had adapted centuries-old irrigation infrastructures to new imperial purposes with impressive results. To the

American engineers who traveled to see these projects, monumental technology and staggering efficiency served as both example and challenge.[91]

As Worster has argued, American engineers' self-comparison with their British counterparts was neither coincidental nor unimportant. The scientific control of water in the West, he maintains, was part and parcel of the American imperial expansion at the turn of the twentieth century. Although John Wesley Powell, the famed Western explorer and early director of the U.S. Geological Survey, was a vocal advocate for the irrigation-as-democracy ideal, his calls for extensive scientific surveys to support environmentally responsive forms of technology and government fell on deaf ears within a Congress that was more focused on capitalist expansion for imperial aims. Popular sentiment focused primarily on the imperial promise of irrigation, as reflected in the broad circulation of books like William Ellsworth Smythe's *Conquest of Arid America* and journals like his *Irrigation Age*.[92]

> The American empire lay waiting in the western desert, and with the irrigated produce from it Americans could go overseas as agricultural merchants, opening up markets throughout the world, opening up the hungry, insatiable markets of China and India, winning through trade what the Europeans must win through bloody arms.[93]

The benign American empire would be based on rational-scientific decision-making entrusted to a technocratic elite through the 1902 Reclamation Act and the centralized authority of the State Engineers it spawned throughout the West.

Against this backdrop, Lowell's vision of an irrigated Mars hardly seems outrageous. On arid, imperiled Mars, the best and brightest were entrusted with responsibility for massive engineering works that would provide the greatest good for the greatest number. In that sense, the newly created U.S. Reclamation Service would have felt right at home on the red planet. Lowell's prose typically included direct comparisons to the American West to make the similarities clear for his popular audiences. Describing the most likely response of Martians to increasing aridity, for example, he conjured a scenario that implicitly endorsed the American reclamation policy as the proper response of intelligent civilizations to arid environmental conditions: "In a word, irrigation on a stupendous scale would be attempted, involving not merely the few hundred thousand acres of an American alkali desert, but an entire planet."[94] Lowell's Bostonian ally Morse likewise conjured history's "great irrigating works . . . from the rude irrigating canals of ancient Peru and Arizona to the marvellous accomplishments of the hydraulic engineer in India and Egypt" in his defense of Lowell's hypothesis, relying on public familiarity with irrigation to dismiss skeptics' arguments about the invisibility of such infrastructure from Earth:

> To realize the extent of this work, it is only necessary to state that in Egypt 6,000,000 acres depend upon irrigation, and this area to be vastly increased in a short time; the Western states of America with 10,000,000 acres, and this area being rapidly augmented by the work of the United States Reclamation Bureau; in India 25,000,000 acres under irrigation, and this being continually added to.[95]

In a different historical era or cultural setting, perhaps the geometric maps of Mars would have conjured other explanations. For Lowell's popular audiences, however, all attention was fixated on the supposed dearth of Martian water in the "dreary wastes of desert land."[96]

As an evolutionary wonderland of advanced technology and peaceful social relations, then, Lowell's Mars stood as an example and beacon for the Western world. Providing an example of Earth's likely future, Lowell argued that Mars should provide hope for those distressed by contemporary concerns such as the management of finite natural resources, the intricacies of American entry into global trade, and domestic class warfare. "In the Martian mind," he commented, "there would be one question perpetually paramount to all the local labor, women's suffrage, and Eastern questions put together—the water question. How to procure water enough to support life would be the great communal problem of the day."[97] In an argument that was mentioned repeatedly in the popular press, Lowell stated that the complexity of the Martian irrigation system proved its builders had advanced beyond the need for petty squabbling and warfare. Upon the evidence of the global canal network, he pronounced that the red planet must be a utopia of sorts:

> Girdling their globe and stretching from pole to pole, the Martian canal system not only embraces their whole world, but is an organized entity. . . . The first thing that is forced on us in conclusion is the necessarily intelligent and non-bellicose character of the community which could thus act as a unit throughout its globe.[98]

As the importance of resource concerns began to outweigh political affairs, Lowell asserted, the evolutionary process would necessarily lead to worldwide peace:

War is a survival among us from savage times and affects now chiefly the boyish and unthinking element of the nation. The wisest realize that there are better ways for practicing heroism and other and more certain ends of insuring the survival of the fittest. . . . Whether increasing common sense or increasing necessity was the spur that drove the Martians to this eminently sagacious state we cannot say, but it is certain that reached it they have, and equally certain that if they had not they must all die. When a planet has attained to the age of advancing decrepitude, and the remnant of its water supply resides simply in its polar caps, these can only be effectively tapped for the benefit of the inhabitants when arctic and equatorial peoples are at one. Difference of policy on the question of the all-important water supply means nothing short of death. Isolated communities cannot there be sufficient unto themselves; they must combine to solidarity or perish.[99]

With each successive publication, Lowell became more and more certain of this pronouncement. In 1903, for instance, he reported a peculiar phenomenon in his observational data. Having regularly mapped a certain canal since 1894, he discovered that it actually seemed to show up in one of two slightly different positions, depending on the year in which his observations had been conducted. On these data, Lowell determined that he must actually be observing two separate, adjacent canals. The importance of this observation, he claimed, was the fact that the two neighboring canals never appeared simultaneously, meaning that one must always be dry while the other was supplied with water. As this phenomenon could not be explained by simple meteorology, Lowell hailed this as an example of a sophisticated water-sharing arrangement implemented by peacefully organized regional neighbors: "It is easily conceivable that a limited water supply should involve a necessity of the sort. It may well be that after one district has enjoyed the water and its results for a certain period, the supply should then be turned for a time into a neighboring one to be turned back again after a while." His interpretation fit perfectly with his theory that water crisis was the source of societal advancement on Mars. Lowell offered this apparently high level of cooperation among neighboring regions as further evidence of the Martians' impressive social organization and advancement, lauding their civilization as an example of the highest level attainable in a Spencerian hierarchy of cultures.[100]

The struggle for existence in their planet's decrepitude and decay would tend to evolve intelligence to cope with circumstances growing momentarily more and more adverse. But, furthermore, the solidarity that the conditions prescribed would conduce to a breadth of understanding sufficient to utilize it. Intercommunication over the whole globe is made not only possible, but obligatory.[101]

It is this idea of environmental change leading to peace between nations and cultures that particularly thrilled the American audiences that flocked to Lowell's lectures on the East Coast and snapped up copies of his books. Describing Lowell's mesmerizing effect, a *Boston Commonwealth* article reported that he had painted a vivid picture of "a population which has not spent, apparently, most of its history in mutual throat cutting and constant quarreling," thus allowing for the construction of "marvels of irrigation and vegetation which we see upon the planet Mars to-day."[102] Likewise, an American reviewer's assessment of Lowell's *Mars and its Canals* found it entirely convincing in suggesting that cooperation was "the distinctive feature of life on Mars:"

[T]he inhabitants of Mars cannot indulge in the practice of war at any spot on their globe. The compelling motive has to do with the necessity for husbanding water. . . . A war on Mars having anything like the aspects of those sanguinary conflicts of which the earth's history is so full would terminate the career of the Martians as effectively as the ravages of the Punic wars led to the destruction of Carthage. Irrigation on Mars is existence.[103]

The Lowellian construction of an irrigated Mars thus supported the view that civilization could be sustained in arid landscapes via science, technology, and centralized social organization, despite any cosmic inevitability of planetary demise. In this regard, the Lowell-inspired discussion about Mars echoed the work of Marsh, who rejected Huntington's climatic determinism and insisted that humans were an active agent in the modification of the Earth. Marsh's worry that humans had developed the capacity to irrevocably alter natural landscapes at a global scale was, paradoxically, the Lowellian Martians' salvation. Mars thus found itself at the intellectual center of one of the key questions in geographic thought. The opposing stances of Huntington and Marsh focused on the relative ability of environments to affect humans and of humans to affect environments. Mars, as an example of both, offered a venue for contestation of the two viewpoints.

The views emanating from Europe, especially Britain, were somewhat different from the American treatment of these same questions. Wallace, for his part, found most of Lowell's assumptions about Martian irrigation to be positively ridiculous, labeling them "wholly erroneous and rationally inconceivable."[104] Wallace dismissively expressed Lowell's explanation of Mars in the following terms:

> [T]he inhabitants of Mars have carried out on their small and naturally inhospitable planet a vast system of irrigation-works, far greater both in its extent, in its utility, and its effect upon their world as a habitation for civilized beings, than anything we have yet done upon our earth, where our destructive agencies are perhaps more prominent than those of an improving and recuperative character.[105]

Wallace thus focused, like Marsh, on the ability of humans to impact their environments, while rejecting Lowell's suggestion that such impacts could be positive, rather than negative.

To a large extent, the Lowell-Wallace disagreement over irrigation reflects not only differing political perspectives but also differing national experiences with both imperialism and water engineering. Europeans had long been interested in irrigation in the imperial context. The famous French survey of Egypt during the 1798–1801 invasion campaign had fascinated European readers with its maps, sketches, and descriptions of a waterless landscape in which civilization depended heavily on its irrigation systems. When the British assumed control of Egypt later in the nineteenth century, they brought engineers from India—where the British colonial administration had established a professional school to train civil engineers in hydraulics and water management—to rehabilitate the Egyptian irrigation system. For many of the European nations, water engineering was central to the colonial project.[106] The European publics were therefore accustomed to news about the digging of canals, the constructing of dams, the draining of swamps, and the pumping of water, especially in the arid landscapes of North Africa and South Asia. (A British astronomer had drawn on this assumed knowledge in an 1892 comment about Martian conditions, well before Lowell entered the discussion: "One requires to have seen an Indian river, or better still, the Nile valley to realize what an inundation may mean, and especially under the conditions which have now been established to exist on Mars.")[107] Unlike the American case, however, where irrigation projects were at least rhetorically justified as a means of expanding democratic society, the European focus on water engineering was oriented quite directly toward environmental and social control. Although visiting American engineers marveled at the high efficiency of British waterworks in India, for instance, other commentators noted the "anti-democratic" tendency toward "despotism" embodied in the ends they served. Both British engineers and publics acknowledged their very impressive irrigation technology as an explicit tool of empire.[108]

Given his beliefs on land nationalization and socialism, it should come as no surprise that Wallace was an outspoken critic of the imperial project. His objections to British imperialism, which he summarized in 1899's *Wonderful Century*, catalogued such negative effects as the environmental degradation caused by the clearing of forests, the introduction of cash crops with low stability and high likelihood of land erosion, and the replication of highly unequal social relations that basically enslaved working populations purely for the benefit of the few who accumulated wealth. Wallace's primary example of these imperial ills was the British empire's irrigated jewel, India. In India, Wallace charged, widespread disease and starvation reflected an inferior government, one that continually increased its burdens on Indians and "led to the most cruel oppression."[109] Most important, Wallace felt that irrigation was used as the unfortunate justification for imperial mismanagement:

> The people of India are industrious, patient, and frugal in the highest degree; and the soil and climate are such that the one thing wanted to ensure good crops and abundance of food is water-storage for irrigation, and absolute permanence of tenure for the cultivator. That we have built costly railways for the benefit of merchants and capitalists, and have spent upon these and upon frontier-wars the money which would have secured water for irrigation wherever wanted, and thus prevented the continued recurrence of famine whenever the rains are deficient, is an evil attendant on our rule which outweighs many of its benefits.[110]

Based on this view of India, Wallace did not see water scarcity as leading to benevolent or peaceful governance. Instead, he found it more likely to produce conflict or to be used in the exploitation of those already oppressed.

Wallace's rejection of Lowell's irrigated-Mars hypothesis clearly bears a direct relationship to these concerns. Some of his most bitter attacks were reserved not for the improbability of Lowell's celebrated Martians but for the improbability of their supposed irrigation infrastructure. The fact of water scarcity, Wallace argued, would logically preclude the intraregional or global cooperation that Lowell assumed to be its most likely result. Drawing heavily on an earlier critical publication by the popular British astronomy writer Agnes Clerke, Wallace faulted Lowell for his social interpretation of irrigation:

[H]e never even attempts to explain how the Martians could have lived *before* this great system was planned and executed, or why they did not *first* utilize and render fertile the belt of land adjacent to the limits of the polar snows—why the method of irrigation did not, as with all human arts, begin gradually at home, with terraces and channels to irrigate the land close to the source of the water.[111]

As Clerke had argued in her original article, "it would not be left to their discretion to share with the opposite hemisphere supplies which would certainly fall short of what was wanted for their own." In Wallace's view, such water-hoarding and its associated conflicts would only be exacerbated as supplies dwindled. Given that he was convinced Mars had a "totally inadequate

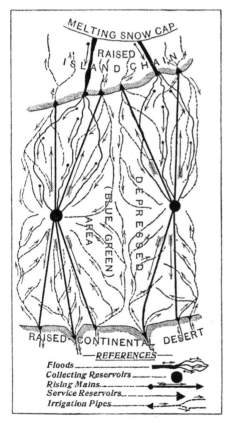

Fig. 2.—The irrigation of a depressed area in Mars's southern hemisphere.

Figure 5.9. Illustration of Housden's hydraulic conception of Mars, 1914. This image was published in *Scientific American*'s review of the book *Riddle of Mars* in the August 15, 1914, supplement (on p.106). It shows Charles Housden's hydraulic conception of the engineered movement of water from Martian poles to equatorial regions.

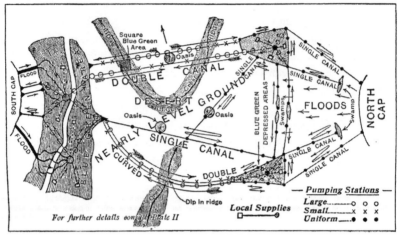

Fig. 3.—Plan illustrating the transport of water across Mars.

Figure 5.10. Illustration of Housden's irrigation scheme for Mars, 1914. This image was published alongside fig. 5.9 in *Scientific American*'s review of the book *Riddle of Mars* in the August 15, 1914, supplement (on p. 106). It shows author Housden's detailed conception of the hydraulic engineering necessary to move Martian water at a global scale.

water-supply for such world-wide irrigation," he could only reject Lowell's predicted cooperative irrigation scheme.[112]

Wallace supported these key points by charging that Lowell's system would also be supremely inefficient and therefore fundamentally improbable: "he never even discusses . . . the extreme irrationality of constructing so vast a canal-system the waste of which, by evaporation, when exposed to such desert conditions as he himself describes, would use up ten times the probable supply." Although a retired British colonial irrigation engineer who also happened to be a member of the British Astronomical Association was moved to publish a book-length rebuttal of these inefficiency charges (*Scientific American*'s review of this volume included the illustrations in figs. 5.9 and 5.10), Wallace's social assessment of irrigation and its ills was never addressed by Lowell or any of his allies. Even in his 1908 *Abode of Life*, in which Lowell summarized his Mars hypothesis one final time, he did not respond to this part of Wallace's critique other than to clarify that "The thing was not done in a day. . . . Probably the beginnings were small and inconspicuous, as the water at first locally gave out. From this it was a step to greater distances, until necessity lured them even to the pole." In this and other reassertions of the Martians' intellectual advancement, however,

Lowell never came up with a direct explanation for how global cooperation could logically evolve from desperate water scarcity.[113]

Conclusions

In Lowell's and Wallace's competing perspectives on irrigation, we see the reflections of their divergent views on the human–environment relationship. In Lowell's Progressive view, environmentally driven advances in human intelligence and technology paradoxically allowed for massive control over and alteration of the originally decisive environments. In Wallace's socialist view, on the other hand, the interaction of humans with arid environments tended to produce only conflict and degradation of both environments and societies. In this, he was followed by Wittfogel, whose now infamous Marxist-derived theory of hydraulic civilization focused almost exclusively on the tendency toward despotism in irrigation-dependent societies.[114] For all the fundamental differences between these perspectives, they both included critiques of environmental determinism, albeit in divergent and competing ways. Lowell's insistence on the fundamental power of Martian irrigators to alter their planet (and their survival prospects) via massive engineering and technology, for instance, is not so far from Marsh's concern that humans were agents of environmental destruction. Although Marsh was fundamentally concerned with the negative environmental impacts of human action, Lowell's optimism certainly relied on the same general principle: that humans are not fully controlled by their environments, however important they may be in determining the nature of civilization's progress. Wallace's insistence that cooperative Martian irrigation would be impossible at a global scale likewise rejected environmental determinism. In its place, he offered a historically deterministic view that viewed social, economic, and political influences as much more influential than environmental constraints. These public intellectuals' competing views on the Mars question reveal its deep implication in philosophical debates and help explain both the seriousness with which it was treated and the enthusiasm with which it was received.

SIX

Toward a Cultural Geography of Mars: Imaginative Geography and the Superior Martian

Those who have never seen a living Martian can scarcely imagine the strange horror of its appearance. . . . Even at this first encounter, this first glimpse, I was overcome with disgust and dread.

—H. G. Wells, *War of the Worlds* (1898)

From debates over the nature of Martian landscapes and nature–society relations, it was only a small step to consider the characteristics of the supposed Martian irrigators themselves. Although very few astronomers would allow themselves to be drawn on the issue of Martian physique or intellect, popular writers engaged in active speculation as to what the Martians were like. Did they have wings? Could they speak English? Did they know how to control electricity? Although some of the writing and imagery in this vein was clearly meant to be fanciful, the discussion as a whole was dominated by serious themes that both reflected and challenged familiar geographical ways of thinking.

In an 1896 *Harper's* "Editor's Study," for instance, Charles Warner reckoned that the canal-builders of the red planet would likely appear bizarre to human observers.

The forms may be in the shape of a single wheel, able to roll easily anywhere, or in that of a sphere, or of a cigar. . . . The Martian lord of creation may have wings, he may be a gigantic insect, or a noble sort of eagle. . . . He may, indeed, have four dimensions instead of three, and instead of five senses a dozen, and among them common-sense. He may be able to see himself as others see him. All his conditions are probably totally different from ours.

Our vices may be his virtues—but we trust not; or he may be without more morals, or even conscience, than a tree.[1]

In this characterization, as in others that appeared in broad-circulation publications around the same time, the Martians were depicted as classic "Others": strange and fascinating opposites who were nonetheless comprehensible through analogical comparison of their limited similarities. For the case of the Martian Others, differences were typically stated in physical and intellectual terms while analogies of similarity were found in the Martians' presumed ability to modify landscapes through irrigation.

As much as these narratives followed the typical geographer's formula for representing non-Westerners, however, they also present a serious obstacle for any explanation of Mars science as geographical science. European and American representations of non-Westerners at the turn of the century were typically defined by characterization of Others as inferior; Martian civilization and intellect, by contrast, were consistently constructed as *more* advanced and complex than those on Earth. The construction of both bewildering Otherness and technological superiority for Martian beings thus challenges what might otherwise be expected from a geographically inflected discourse about foreign landscapes and peoples. The bulk of this chapter explores the nature and meanings of this challenge, focusing on contrasts in the ways that British and American audiences grappled with the idea of Martian superiority and Otherness. The concluding section puts this challenge in the context of Mars science's broad geographical relevance at the turn of the twentieth century, taking up strands of argument that have been presented throughout the book.

Imaginative Geography

Speculating about and defining the characteristics of strange Martian beings was not a mere diversion for the Western audiences that consumed astronomical science through its popular interpretations. Westerners have long defined themselves in relation to strange Others, and sociologists have consequently argued that the rise of conceptual categories such as "the West" and "Westerners" was inextricably linked to the discursive definition of non-Westerners as inferior or subject to Westerners. These divisions have generally reinforced the hegemonic superiority of the West and disadvantaged its Others within the particular constellation of social, political, and economic arrangements that have come to define modernity.[2]

At the time of the Mars sensation, there were some well-defined discourses about non-Western Others already in active operation. According to Edward Said's now-classic argument regarding "Orientalism," for example, European (especially French) scholars had engaged since the Enlightenment in textual representations of the Islamic world that consistently presented its landscapes and peoples as similar, but inferior, to Europe in a multitude of ways.[3] Drawing heavily on Foucault's arguments about the relationship between knowledge and power, Said argued that Europe's long interaction with the Islamic world from a position of power was achieved through the process of "imaginative geography"—the discursive construction of geographical knowledge through uncritical repetition of simplistic yet powerful tropes and analogies. As defined by Said, an imaginative geography has little to do with the actual physical or cultural geography of a region, but it nonetheless functions as a powerful influence and constraint on the subsequent production and ordering of geographic knowledge for that region. Although Said's influential argument has undergone fairly rigorous critique in the decades since it was first presented, recent work has confirmed the usefulness of the imaginative geography concept in understanding the processes by which Western societies, governments, institutions, and individuals have defined themselves.[4]

As traced through certain tropes and literary conventions—including narrative voice, literary structure, figures of speech, images, themes, and motifs—Said had identified a common "set of representative figures" in Western writing about the Islamic world.[5] As much post-Said scholarship has now shown, these essential elements have also appeared in numerous other imaginative geographies beyond the French/European representation of the Islamic world. Motifs of backwardness, depravity, and the conflation of Others with nature have gone hand in hand with tropes of the noble savage, the lost paradise, and the superiority of European civilization through technology.[6]

To a large extent, the turn-of-the-century writing about Martians followed these conventions of Orientalist imaginative geography. Writers typically cast the planet as alien and impenetrable while at the same time familiar and open to a scientific gaze. Despite the prevalence of analogical reasoning and comparison in representation of Mars, the popular sensation also trafficked very heavily in narratives of environmental and cultural difference that cast Mars as fundamentally strange. The environmental features most commonly cited to illustrate the strangeness of the Martian environment were the Martian moons. With two moons, both of which were much smaller and closer to the planet's surface than Earth's moon, Mars was said to enjoy a spectacular night sky, in which one of its two moons was almost always visible. The nearest and most prominent moon (named Phobos), in fact, revolves around its parent planet so rapidly that it sometimes rises twice in a single night, during which a hypothetical Martian observer would see it pass through all its lunar phases in a quick passage from west to east. The fact that Phobos rose in the Martian west and set in its east was mentioned again and again in popular reports as "a phenomenon hitherto unknown in the solar system," underscoring a fascination with the differences between Earth and Mars.[7]

The climate and surface hydrology of Mars likewise fed into a narrative of environmental difference, with speculations about Martian weather, or lack thereof, cropping up particularly frequently in the popular press. Spurred by astronomers' back-and-forth debates about the density and thickness of the Martian atmosphere, mainstream writers often commented on Martian climate. Though Lowell and his allies were most concerned with proving the existence of an atmosphere dense enough to support intelligent life, popular audiences tended to focus on comments about how this atmosphere behaved. Seizing on Lowell's suggestion that the extreme rarity of the Martian atmosphere would probably modify, but not prevent, its functioning in support of plant and animal life, commentators marveled at the idea Mars could be free from storms. If Lowell was right in postulating frost as the most common form of precipitation on Mars, for example, the standard terrestrial experience of a rainstorm would be nonexistent on Mars. A Martian snowstorm, according to one magazine, would likewise be "not a howling, blinding gale accompanied by irresistible snowdrifts, but, at most, a soft zephyr attended by a gentle and almost impalpable deposit of snow."[8] The idea that Mars might be liberated from one of the primary constraints on humans' everyday life—inclement weather—clearly caught the attention of numerous audiences. The perplexing strangeness of an environment that experienced little more than gentle breezes and occasional frosts drew significant attention to the Martian hydrologic issues that Lowell set forth as central to his inhabited-Mars hypothesis. He apparently recognized the value of turning attention to weather issues early, as one of his first *Atlantic Monthly* articles included quips later repeated in his first broad-circulation book *Mars*: "Mars is blissfully destitute of weather. Unlike New England, which has more than it can accommodate, Mars has none of the article. What takes its place there, as the staple topic of conversation for empty-headed folk, remains one of the Martian mysteries yet to be solved."[9] In the popular sensation that developed around Lowell's hypothesis, topics like these drew significant attention. Where scientists tended to concentrate on technical points of analogy and difference, popular audiences took an equal interest in such phenomenological reasoning.

Conditioned by this narrative of environmental difference, the popular sensation also embraced questions of how Martian peoples might be different from humans. Acknowledgments that Mars' landscape was not entirely like Earth's produced a wide variety of speculations over what the Martians might be like. These ranged from physiological speculations over Martian bodily form to utopian ponderings about the likelihood that Martians had achieved "peace [and] halcyon spells." In general, Martians were presumed to be physically different from humans, although their cultural and moral characteristics were always left much open to debate. Lowell himself was again responsible for initiating the fascination over Martian bodies, as he coyly asserted that "we lack the data even to conceive" what manner of beings might have developed in "surroundings so different from our own." Although he frequently made statements like this to deflect criticism that he was engaging in unscientific speculation, popular writers regularly repeated his assertion that there was no way to know what the Martians were like. As interpreted in nonspecialist contexts, this clearly meant that Martians must be very different. Most commonly, Martian beings were presumed to be "probably bigger, owing to the reduced gravity of Mars."[10] If bigger, then they could presumably also perform more work.

> Assuming that the Martian is a creature thus constructed and regarding him merely as a machine with brains, his canal excavating possibilities, on a planet where bodies weigh only one-third as much as on the Earth, become truly awesome. A Martian laborer could perform as much work in a given time as fifty or sixty terrestrial ditch diggers, and keep pace with a powerful Panama dredger.[11]

It was only on this point, in fact, that Lowell made explicit arguments about Martian physical stature, arguing that the reduced-gravity environment on Mars would necessarily improve any inhabitants' ability to perform feats of strength. This argument, of course, identified a mechanism by which the hypothetical Martians would have been able to build an irrigation network so massive that it seemed impossible to many on Earth. The sensational interest in Martian bodies and strength, therefore, led audiences to focus on a topic that supported Lowell's hypothesis: that visible Martian landscape patterns should be interpreted as evidence of a massive global irrigation network.

Newspapers followed this example in their sensational writing on Mars, often pointing out how strange a place Mars would seem to any human who managed to visit, as in this list of things "men of the earth might do on Mars": "A baseball player could knock a baseball the distance from the Battery to City Hall. A gunner handling a 16-inch gun could shoot the distance from New York to Poughkeepsie. . . . Men could run almost as fast as horses run on the earth. In case of fires men might jump from third-story windows without injury."[12] As Warner claimed of the Martian in his *Harper's* article: "All his conditions are probably totally different from ours. Our vices may be his virtues."[13]

In these terms, Mars sounds like the classic antipode—a place of utter opposites in every dimension. At the same time, however, the appearance of detailed descriptions of Mars implied that astronomers had conceptual control of it. Every scientific report—even those emphasizing its perplexing strangeness—indicated that Mars was open to scrutiny. Much was made of the fact that Mars had no significant clouds and was therefore visible in its entirety, unlike cloudy Earth, which would appear shrouded to a hypothetical outside observer. Laid bare for the outside observer, Mars was presented as if it were completely accessible, despite its mesmerizing peculiarities. The fact that astronomers used impressive telescopes to effect a quintessential all-seeing scientific gaze only served to reinforce the power of this rhetoric.

These constructions of Mars functioned as an imaginative geography in ways very similar to those Said had identified as emerging from Europe to represent and control the Orient. Through simplistic tropes implying both familiarity and difference and through the use of a rhetoric of scientific objectivity, claims about Mars became entrenched through regular repetition in the popular press. The uncritical use of terrestrial vocabulary such as "polar caps," "canals," and "oases" began innocuously enough in reference to purely visual similarities. By the time the public began to take interest in the topic of Mars, however, such terms had been repeated so many times that they had already begun to produce a powerful imaginative geography of the red planet as having substantial amounts of water. As one early opponent of the inhabited-Mars hypothesis pointed out, the word "canal" itself conjured images that were unprovable yet extremely powerful:

> It has been previously remarked that the name of "canals" as applied to the dark streaks on Mars . . . does not really matter, the term being merely used in a technical way, and not as implying that the so-called canals are artificial productions or even water at all, and that the terms seas and bays are merely convenient ways of referring to these details, and do not in any way prejudge

the question. I am afraid the explanation does not suffice to remove the impression. You cannot constantly allude to a man as "that nigger," however much you explain that you do not mean to imply anything about his colour, without creating an impression that he lacks something of desirable whiteness. . . . I fear the direction given to observation and deduction by the tacit assumption that the phenomena of Mars are due to water, all water and nothing but water, is somewhat injurious.[14]

Just as occurred in essentialized representations of non-Europeans, then, such simplifications of terminology allowed diverse audiences to conceive of Mars as a familiar world while also promoting a very specific view of its geography.

Superior Martians and the Reverse Gaze

In a critical inversion of the dominant trends in imaginative geography, however, the observed Martians were thought to be greatly superior to the Western astronomers and audiences who observed them. This idea was first spurred by early comments regarding Martians' ability to observe the Earth, which ironically began in the cautious pages of astronomical journals and textbooks. From the earliest astronomical reports on Martian surface features, astronomers had often compared their own views of Mars with the hypothetical view a Martian would have of the Earth. The famous English astronomer Herschel suggested, for instance, that the reddish color of Mars indicated "an ochrey tinge in the general soil, like what the red sandstone districts on the Earth may possibly offer to the inhabitants of Mars, only more decided."[15] This reference to Earth's visibility from Mars, which was commonly used as a comparison device in the astronomical literature, functioned similarly to the skeptical use of the word "canal" to describe the Martian landscape: even though astronomers typically dismissed the possibility (or at least certainty) of Martian inhabitants, their use of the view-from-Mars mechanism actually reinforced the idea of an inhabited Mars. From this rhetorical foundation, a powerful gazing-Martian trope developed in the popular press and other popular media formats, where it was usually used to paint the hypothetical Martian as an intelligent, scientific astronomer, capable of casting a penetrating reverse gaze toward the Earth (see fig. 6.1). Scores of surviving cartoons and jokes based on the gazing Martian indicate that popular audiences reacted strongly to the idea that Martians might know much more about humans than Earth's own astronomers knew about Mars. In this comment in a popular astronomy magazine, for example,

Figure 6.1. "A Signal from Mars" sheet-music cover, 1901. In this unsigned artwork used as a sheet-music cover illustration, Martians dressed as classical Greek scholars illuminate Earth's surface with a signaling device. Note that the earth itself is shown with latitude and longitude markings, mimicking the graticule shown on the reference globe (and presumably also on the rolled maps) in front of the gazing astronomers. © The British Library Board (shelfmark h. 3286. v.[18]).

the author was rather explicit in ruminating on the intelligence and vision of the Martians:

These facts . . . lead us to speculate as to the kind of inhabitants there may be upon that far-away world, and what they are doing; whether they are like ourselves. Are they devoted to science? Are they constructing immense telescopes and gazing at us, making maps of the Atlantic and Pacific Oceans and the eastern and western continents? Do they know whether, at the north pole of the earth, there is an open polar sea, or whether there is an undiscovered continent near the south pole? Are they a race of great engineers, and do they construct public works on a gigantic scale?[16]

This and other similar comments credited the Martians with impressive capabilities in science, cartography, engineering, and other activities that defined the focus and ambition of the Western powers.

By the turn of the century, the trope of the watchful Martian had become so prevalent that it was regularly used as a lampoon device. Cartoons sometimes speculated on the wild images of Earth that would appear in a Martian telescope. Newspapers filled extra space on their pages with quips about Martians astronomers, such as this example: "A telegram from Prof Lowell at the Flagstaff observatory says that the canals of Mars have been photographed by Lampland. We wonder if Mars is photographing our Panama Canal."[17] Even the more scientifically oriented astronomy journals occasionally participated in such humor. After an unseasonably late snowfall in England in 1908, for instance, the British astronomy journal *The Observatory* included a spoof story supposedly reprinted from the fictitious *Mars Wireless Intelligencer*: "Professor Highell, of Bannerpole, has observed on Terra a brightening of the tiny spot known as Albion, suggesting a fall of snow." (The reference to "Highell, of Bannerpole" is, of course, an allegorical reference to Lowell, who worked at Flagstaff.)[18] These amusements not only poked fun at the uncertainty English and American astronomers expressed in their differing interpretations of Mars, but they also reflected growing comfort with the idea that the red planet might host intelligent beings. It seems to have been commonly accepted that if such beings existed, they must surely be looking at the Earth (fig. 6.2).

Despite the fact that it turned up so frequently in representations meant to be more humorous than serious, the trope of the gazing Martian presents us with an important paradox. Given that the larger scientific narrative about Martian geography was based in so many ways on geographical themes related to imperial and colonial interests, any discussion of Martian cultural geography might be expected to reflect dominant discourses regarding the inferiority, weakness, or backwardness of the inhabitants. Instead, however, the popular narratives of an intelligent, cultured Martian departed in fundamental ways. Unlike the Orientalist propensity to erase existing cultures from a foreign landscape, both textually and cartographically, the Mars narrative paradoxically projected unseen inhabitants.[19] The map of Mars, in

—Professor Lowell's theory of life on Mars inspires the wonder how long

Martians have been making moving pictures of life on earth.

Figure 6.2. Cartoon showing Martian observer of Earth, 1907. This cartoon published in the *New York Times* shows a Martian using a video camera to record images of the planet Earth. Courtesy Lowell Observatory Archives.

fact, was once said to be "too full"—with "none of the tantalizing blank spaces" that exist on Earth's map.[20] Regardless of the fact that Western astronomers had actually constructed the "fullness" of this map in a blatantly territorial struggle, popular audiences no longer considered Mars a conquerable or available territory after 1900. Astronomers built their prestige by overacknowledging the perceived Martian presence in their canal-filled maps, not by conceptually minimizing it.

Even more significantly, the Martians filling the map were widely said to be superior to humans. Popular writers seized on the idea of a superior Martian, regularly characterizing Martians in terms of their advantages over humans. In a *Cosmopolitan* article ostensibly limited to logical scientific conjectures about the Martians' characteristics, for instance, Wells claimed, "The Martians are probably far more intellectual than men and more scientific, and beside their history the civilization of humanity is a thing of yesterday."[21] Such statements were ubiquitous in the popular press, providing a sharp contrast to the statements traditionally mobilized to characterize "savages" or "natives." Although imperial-era Western writing sometimes praised the civilizations of the Other, compliments were generally focused on mere exotic charm or on a long-lost, glorious classical past. Present-day occupants of non-Western lands were simply never characterized as "far more intellectual and scientific" than Westerners in the dominant discourses of the day.

Signaling and Subjectivity

In addition, the imagination of a reverse gaze attributed remarkable abilities of self-representation to the supposed Martians. Popular audiences who accepted the inhabited-Mars hypothesis typically assumed that the advanced Martians had been hard at work not only digging colossal canals but also trying to communicate with people on Earth.[22] Beginning in the 1890s, astronomers occasionally reported seeing "projections" on the dark side of Mars: small bright markings that occurred very near the "terminator" edge, or the line where sunset was falling on the Martian surface (see fig. 6.3). Because of the markings' invariable proximity to the sunlit half of the planet, astronomers assumed that the bright spots were likely to be high clouds or mountaintops illuminated by the lingering twilight sun even though the surrounding areas were already shrouded in darkness. Astronomers' initial uncertainty about this interpretation, however, allowed room for an altogether different line of speculation. Following several telegrams sent by astronomers to the newspapers in 1892, an explosion of press coverage established the idea that fleeting bright markings on the Martian surface were likely to be light-beams deliberately flashed as signals to the Earth.

Astronomers were rattled by the sensationalism with which this storyline gathered pace but found themselves almost powerless to stop its spread. A note in the *Journal for the British Astronomical Association* could barely contain its disgust:

> If one may judge from the telegrams and articles which have appeared in the newspapers with regard to the present Opposition of Mars, astronomy is making great progress in popular interest, though much that has been written shows that the public still require further education. To begin with, it was evidently expected that Mars would do something extraordinary,—flash a congratulatory communication by the Morse code at least—to celebrate his coming successfully out of opposition. Then our American colleagues suffered

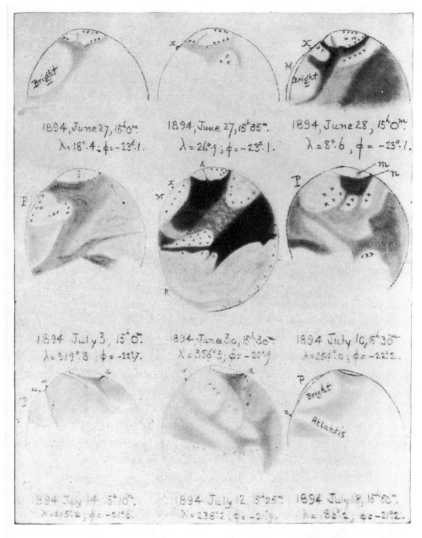

DIAGRAMS OF MARS, SHOWING BRIGHT PROMINENCES.

By Edward S. Holden.

Figure 6.3. Diagram showing bright "prominences" on Mars, 1894. Lick Observatory astronomer Edward Holden published these sketches in the December 1894 number of the *Publications of the Astronomical Society of the Pacific* (facing p. 286). They record the locations of bright "prominences" observed on the surface or in the atmosphere of Mars.

many things of many reporters, and it is to be feared (or rather hoped) were grievously slandered by them. . . . [T]he leading idea in several papers seemed to be the prospect of starting an interplanetary telegraph.[23]

Despite this British protest, the damage had already been done. The Mars craze had officially opened in America, spurred by the idea that an intelligent race of Martians was trying to communicate with humans.

By century's end, newspapers and popular magazines were regularly reporting on Mars, including a broadened discussion of the possibilities of signaling between the two planets. Regular reports of "terminator projections," especially from Lowell Observatory, continued to fire speculations about the Martians' use of electrical light-beams (see fig. 6.4). Though Lowell himself classified the projection phenomena as clouds, his insistence that Mars hosted intelligent life clearly contributed to the popular belief that they could be signals.[24] The famous engineer and inventor Nicola Tesla fanned the flames in 1900 by reporting that he had detected an odd electrical transmission in his Colorado mountaintop laboratory. Claiming it could not be explained by the well-known effects of the sun, the aurora borealis, or the Earth, he determined it had likely come from Mars. The supposed message—"one, two, three"—not only confirmed a Martian knowledge of mathematics, he said, but also called for a response from humans. "Absolute certitude as to the receipt and interchange of messages would be reached as soon as we could respond with the number 'four,'" he claimed. "The Martians, or the inhabitants of whatever planet had signalled to us, would understand at once that we had caught their message across the gulf of space and sent back a response."[25]

Despite continued attempts by leading astronomers to divert attention from such statements, the British and especially American publics seem to have been captivated by the idea of sending a return signal to Mars.[26] As Lowell's hypothesis about hyperintelligent, canal-digging Martians became more entrenched, the desire to "enter into telephotic communication" as a means of understanding the nature of Martian advancement seemed more and more urgent.[27] Interplanetary communication emerged as a topic in poetry, popular music, cartoons, and even a well-received play staged in London.[28] Literary essays discussed the significance of human–Martian communication, while technical articles debated the means by which it could be achieved. Proposals for how to reply to incoming signals—which included raising a huge flag, carving an enormous message into the Saharan landscape, reflecting daytime sunlight via mirrors, systematically turning electric lights on and off in an urban area at night, or even using wireless

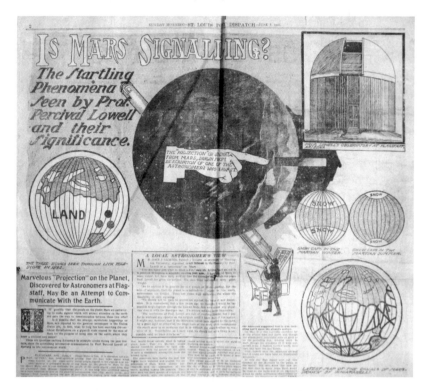

Figure 6.4. Newspaper report on Martian signaling, 1903. This feature story, published in the June 7, 1903, *St. Louis Post-Dispatch*, shows that interest in Martian signaling persisted into the early twentieth century. Courtesy Lowell Observatory Archives.

telegraphy equipment carried in hot-air balloons—fascinated audiences well into the first decade of the 1900s.[29] Cautions about the immense difficulties and costs of implementing such proposals did not deter many enthusiasts, as reflected in this 1909 magazine commentary: "[S]upposing we kept up the signals at every opposition until the next equally favorable one, fifteen years from now, and then received their answer, would it not pay? Would not the achievement be momentous?"[30] In fact, the popular discussion of Martian signaling soon began to treat it as a technically simple issue that required little more than political will. Tesla claimed that he had already devised a method of communication but was simply too busy with other projects to put it into place. Perhaps heeding the tone of this commentary, a donor to the French Academy of Sciences bequeathed "the sum of one hundred thousand francs to him who should find the means of communicating with a star, excepted Mars; doubtless because, in her mind, it would be too easily and too quickly done."[31]

As might be expected, astronomers actively rejected most of this speculation as sensational and impossible, but their repeated dismissals did not diminish popular interest in the possibility of communication with Mars. The enthusiastic popular consideration of signaling as both feasible and desirable points toward an important assumption that was never addressed by specialist astronomers: that there was something important to be communicated between the two planets in the first place. As revealed in the pages of literary magazines and newspapers as well as on the stages of musical and theatrical performers, popular audiences commonly accepted the idea that radically different Martians must surely have some wisdom to dispense for the benefits of Earth's inhabitants (since no one claimed that humans had anything of import to say to the Martians).

As compared to the supposed lack of subjectivity afforded to the Other in most Orientalist writing, the Martians' assumed representational ability was astonishing. The signal-sending inhabitants of the red planet not only knew beyond a shadow of a doubt that Earth hosted intelligent beings, but had also devised a way of making their presence known across millions of miles of space. The laughable observers on Earth, by contrast, could not even agree as to whether they were witnessing signals or natural phenomena. And even if they were signals, humans could not envision a plausible way to send a response or determine what to say in reply. The Martian Other thus outranked the Western astronomer and his audiences in terms of subjectivity, technology, intelligence, and organization.

Perspectives from Imperial Britain

In looking more carefully at the ubiquitous trope of Martian superiority, however, we see that not all writers, artists, and audiences engaged with the idea in the same way. In general, Lowell's American audiences embraced the prospect of advanced Martians, while Wallace's British countrymen had a much more hesitant response. These contrasting audience reactions can be traced primarily to the varying understandings of cultural geography that had emerged in different national contexts and from different imperial experiences. In making this argument, I depart significantly from Markley's stance on the same issue, in which he sees the reverse gaze primarily as a means of ecological (rather than cultural-geographic) theorizing.[32] Without rejecting his argument that the seeds of Mars' importance to ecological thinking were sown in the early twentieth century, evidence presented here shows that cultural concerns were dominant within early versions of the gazing-Martian construction. Given that imperial activities supported (and demanded) much of the geographic theorizing of the day, we know that British and American geographers pursued different research and pedagogical agendas.[33] As the red planet's commentators turned to geographical literature in their quest to understand theoretically superior Martians, the influence of these divergent imperial contexts produced divergent audience responses.

British audiences, in general, had a very ambivalent response to the construction of the superior Martian. This stemmed partly from the British astronomical establishment's long denial of the existence of Martian canals, starting with the nationalistic Schiaparelli-Green cartographic debates of the 1870s. By the time Lowell entered the scene in the 1890s, he found a number of British astronomers to be his most vocal critics. In the British popular press, however, writers expressed more interest and less hostility to Lowell and his theories. Although British newspapers often printed Lowell's reports with a critical spin that acknowledged British astronomers' vigorous skepticism, they were not averse to engaging directly with the issue of Martian characteristics. In magazines, reviews, and fiction, particularly, British writers grappled with the implications of Martian technological superiority, often betraying pronounced misgivings.

In one of two main lines of response, British writers typically assumed that any superior Martian beings would necessarily be somewhat like the British themselves: seated at the pinnacle of civilization, with unprecedented technological and social control over environments and peoples. Even if the Martians had managed to harness energy on a grander scale or to achieve a more penetrating interplanetary gaze than that available to fog-bound London astronomers, many writers assumed the inhabitants of the red planet were surely only a step or two removed from Earth's own self-perceived superior race.

The assumption of fundamental likeness, however, presented some obstacles to full acceptance of the superior Martian, primarily because the British had a hard time actually seeing themselves in the Martian landscape. First, the irrigation scheme didn't quite fit British expectations. Though general audiences weren't unduly swayed by Wallace's argument that extreme resource scarcity could never lead to global cooperation, there was still the issue of geometry. None of the British irrigation systems looked remotely like the spoke-like organization observed on Mars. As one commentator noted, it was hard to imagine an artificial landscape that wouldn't follow natural patterns in the landscape, such as those related to the British-controlled irrigation of the Nile Valley.

> Common sense will suggest that we conform as far as possible to the courses which nature has suggested. . . . We have just one example of a long narrow strip of irrigation—the Nile valley—which might give to an observer upon Venus something of the effect of a Martian canal. . . . But in the Nile Valley the straightness is natural, the narrowness of the fertile strip is natural; and the difference between the old natural and the new artificially-controlled irrigation is principally this, that the irrigation department can store water behind its dams, and supply it at an unnatural time, to allow two crops a year, or to make it possible to grow valuable cotton in place of rice. . . . Mr. Lowell has altogether overdone his insistence upon the artificiality. . . . It is more reasonable to suppose that an irrigation system would be so blended with the natural topography of the planet that artificiality was not its most striking characteristic.[34]

And if it was challenging to project the British irrigation experience to Mars, it was even more difficult to imagine the Martian system imported to Earth. As expressed in the same article, Europe couldn't possibly follow the Martian organization:

> Imagine Europe irrigated upon this plan. Paris, Cologne, Berlin, Warsaw, and Moscow lie on a perfectly straight canal, with irrigation extending ten or fifteen miles on each side, of uniform width, and a circular patch fifty or sixty miles across centred on each city. A second straight canal, with the same features, includes Paris, Berne, Munich, Budapest, and Odessa; a third finds Bordeaux, Munich, Warsaw exactly in a line; a fourth perfectly straight canal runs *via* Hamburg, Berlin, Budapest, Belgrade; and so on.[35]

The patent absurdity of such a geometrically organized Europe confounded the assumption of fundamental likeness between the Martians and Earth's own Westerners.

Some British writers expressed additional resistance to the possibility of European-like Martians because it presented almost too depressing an idea. As popular astronomy writer Clerke put it in her dismissal of Lowell's theories:

> [A]fter so many poignant disillusions, amid the wreck of so many passionate hopes, [humanity] is not enamoured with its own destinies to the point

of desiring to impose them as a maximum of happiness upon the universe. Rather, men cherish the vision of other and better worlds, where intelligence, untrammeled by moral disabilities, may have risen to unimaginable heights, and sense and reason alike are dominated by incorrupt will. But it is improbable that the vision can ever be located in any one of the disseminated orbs around us.[36]

The British desire to equate a superior Martian civilization in some ways with their own thus proved challenging. The reluctance to imagine an intelligent Martian civilization that was not sufficiently like the European civilization to serve as either ally or example points to the conflicted response many British audiences had to the idea of superior Martians. By and large, they simply could not see themselves in the Martian landscape and were therefore inclined either to dismiss the whole concept of Martian inhabitants or to reject the premise of their superiority.

In a second main line of response, however, British writers who went ahead and assumed a fundamental difference (rather than likeness) between their own and Martian civilization found themselves grappling with equally challenging ideas. Most prominently through fiction, writers and audiences considered the possible consequences of contact between Martian and human cultures, frequently coming to frightening or at least uncomfortable conclusions. British historian and writer H. G. Wells' famously chilling novel *War of the Worlds*, for instance, presented a terrifying scenario in which immoral and parasitic Martian techno-droids laid indiscriminate waste to the English countryside during a march on London in search of food resources.[37] Other British fiction that probed the scenario of British-Martian contact likewise addressed deep moral and cultural anxieties. As Schroeder has shown, Mars fiction encapsulated fears then emerging from the terrestrial spheres of intercultural contact: that the Martians might be so different that contact with them could corrupt the home culture completely, that contact could induce a loss of moral identity through "separation from one culture without amalgamation into another" (such as befell Kurtz in Joseph Conrad's *Heart of Darkness*), and that contact could produce an imperial backlash or revolt from those colonized.[38]

British fiction thus squarely confronted the possibility that the Martians' supposed technological superiority might actually mask an abject moral inferiority. Somewhat fearful consideration of this possibility was reflective of the larger European experience, in which imperial activities had raised a host of difficult moral issues that emerged in works like Wells's.[39] One of the most infamously unrepentant British imperialists, Cecil Rhodes, directly rejected these misgivings when he reportedly said from his deathbed in 1902 that his only regret was not having conquered enough land: "The world is nearly all parceled out, and what there is left of it is being divided up, conquered, and colonized. To think of these stars that you see overhead at night, these vast worlds which we can never reach. I would annex the planets if I could."[40] Wells's novel provided, however, a dark prediction of what fulfillment of Rhodes' dying wish would actually mean: not only might the British themselves become the targets of imperial expansion, but the blindness of imperial ambition might also strip away any moral imperatives that had once existed. In this context, the superior Martians that emerged in Wells's novel functioned as "sardonically displaced reflections of Britain's imperialistic appetites and ambitions, stripped of their ideological justifications."[41] Even if they were different from Europeans, the Martians would be likely to face the same upsetting concerns confronting imperial Britain: the contrast between imperial colonialism's benevolent power and its corruptive moral effects.

The general response of British audiences to the superior Martian reflected these misapprehensions, either through skepticism that such a high form of intelligence and organization was even possible, given the pitfalls of European imperialism, or through fear that superior Martians might turn the tables on the dominant West. It is somewhat unclear as to whether British audiences truly perceived the imagined Martian as a cultural "Other" or rather as a frightening possible Self. In either case, the imaginative geography of superior Martians that emerged from British scientists, writers, and audiences carried fair amounts of skepticism. Despite the existence of ardent Lowell supporters in the British Isles, the dominant British narratives reflected caution, doubt, and dread in their consideration of intelligent, canal-digging inhabitants on the red planet.[42]

Perspectives from Exceptional America

In the United States, on the other hand, audiences were much more inclined to accept the superior Martian. Lowell spoke regularly to packed lecture halls across the East Coast, received praise-filled reviews, enjoyed hearty book sales, and became a minor celebrity. His biographer and obituaries all noted this popular success, praising Lowell's ability to reach wide audiences. American newspapers printed Lowell's circulars without criticism, ran speculative stories about the intelligent Martians, and published Mars-related illustrations and maps in full-page formats to catch readers' attention (see,

Figure 6.5. Newspaper report on Martian canals, 1911. This feature story, published in the August 27, 1911, *New York Times*, presents Lowell's announcement of a newly discovered canal as evidence of recent engineering activity on Mars. Courtesy Lowell Observatory Archives.

for example, fig. 6.5, and also refer to fig. 2.12 above). American highbrow magazines took the inhabited-Mars hypothesis seriously, and American fiction established a new science-fiction genre to explore its imaginations of distant Mars.

Even many who did not feel that Lowell had proved his case still acknowledged that superior Martians were eminently possible. Pickering, for example, remained committed to a natural explanation for the canals but admitted that the acceptance of an artificial origin would necessarily prove the existence of intelligent life on Mars:

If they are artificial, it is certain that their constructors possess a knowledge of spherical trigonometry, and considerable skill in the mechanical construction of surveying instruments, implying greater intelligence than that possessed by our ancestors a thousand years ago. It is doubtful if our progenitors in the year 900 A.D. could have built a perfectly straight road three thousand miles long, directed to a definite point, even if it had been across level country.[43]

Though no mention was usually made of Martian literature or arts, the red planet's technologies and engineering prowess were imagined to be unfathomably sophisticated. As Lowell himself put it, even the most magnificent American inventions of the modern era were probably far inferior to the Martian technology: "Quite possibly, . . . with them electrophones and kinetoscopes are things of a bygone past, preserved with veneration in museums as relics of the clumsy contrivances of the simple childhood of the race. Certainly what we see hints at the existence of beings who are in advance of, not behind us, in the journey of life."[44] The imagery that emerged to represent Martians in popular media formats often focused on this sophistication (see fig. 6.6).

In the American embrace of this Lowellian vision, we see the acceptance of an enormous power imbalance with relatively little of the skepticism or panic that afflicted British audiences. Where British audiences had a hard time seeing themselves in the geometric regulation of the Martian landscape, American audiences saw on Mars the logical counterpart to their township-range patterning, already dominant across the landscapes of the central plains. Where British audiences worried about the power of superior Martians to inflict harm on Westerners, Americans were more likely to see the Martians as friendly allies or even as conquerable neighbors. Prolific American popular astronomy writer Garrett Serviss captured the excitement many of his countrymen felt about Martian superiority:

The Martian intelligences might look upon us as we look upon monkeys in a menagerie, and their learned doctors might say: "See what we were like once! These creatures have a gleam of our intelligence, and their limbs and sense organs indicate the line of evolution that ours have followed. They even show the germ of some of our most wonderful organs in their growing sensitiveness to electric forces. Give them time, and place them amid our surroundings, and who knows but that they might develop electro-magnetic vision, electro-magnetic hearing and electro-magnetic muscular control? They might even discover the secret of using inter-atomic energy, which has saved us."[45]

" . . . This ray which has the property of disintegrating the ground by breaking up the atoms of the desert sands, has immense inherent The ground, rocks, sands, etc., everything "melts" before it, as snow goes up in steam before an oxyhydrogen flame. . . ."

Figure 6.6. Illustration for a magazine story on Martian engineering, 1916. This image, published in *Electrical Experimenter* (Gernsback, "How the Martian Canals Are Built," on p. 486), shows an advanced technological society that uses electricity as a means of both flight and landscape alteration. The original caption, which is not fully viewable due to the magazine binding, refers to the "immense inherent power" of the Martian electrical technology.

In the American construction of Mars, then, the Other's technological superiority was a trait to look up to, to emulate, and to use as a guide. Writers uncomplicatedly acknowledged that the Martian behind the telescope knew more about Earth than the American or British astronomer could claim to know about Mars, that the Martian canal engineer had achieved unthinkable levels of earth-moving and water-controlling, and that the Martian signal-makers had harnessed technologies that were still in their infancy on Earth.

Even in science fiction, where the most prominent American writer cast Mars in a fairly dystopian light, we see sharp distinctions with the pandemonium Wells had invoked in *War of the Worlds*. In the United States, Edgar Rice Burroughs's serial adventure stories quickly became the most widely read Mars-related fiction, using the planet "Barsoom" as a direct reference to

Lowell's Mars.[46] Consolidated in several full-length novels, Burroughs's stories cast Mars as a planet gripped by environmental crisis, where the scarcity of water had led to crippling interracial and interurban warfare on the red planet. In contrast to Wells's equally dystopian view, however, Burroughs' stories revolved around the exploits of his hero John Carter, a "gentleman of Virginia" and Confederate soldier who had been unexplainably transported to Mars for ten years. In the reduced-gravity environment of Mars, Carter's physical strength was augmented to the extent that he became a feared warrior and desired suitor. In the end, Carter vanquished his foes, got the girl, and helped save the red planet with some good old-fashioned American valor.[47] Despite echoing the dystopian setting of Wells's novel, then, Burroughs's narrative ended with a powerful American providing a happy ending, in which Mars was saved and Earth was never threatened. This of course contrasted sharply with Wells's resolution, in which an unwitting biological virus (rather than any British imperialist) provided salvation for the Earth, while the future of planet Mars remained unresolved.

The remarkable divergence between British and American audiences' and commentators' responses to Lowell's construction of an inhabited Mars suggests that national context played an important role in the way Mars came to be known at the turn of the century. Given the numerous connections between Mars science and geographical science, which was in turn fundamentally linked to imperial expansionist geopolitics, we must address whether the Mars narratives reflect differing attitudes toward imperialism. As expressed by skeptical astronomers, reticent newspaper publishers, and novelists like Wells, the British attitude was one of wariness and uncertainty, perhaps associated with growing concerns over imperialism's success and influence. In the American enthusiasm for superior Martians, on the other hand, we see a faith in science and technology, particularly in their potential for use as expansionist devices of unification and progress, rather than conflict or domination. As Ward's editor commented on his positive interpretation of the certainty of planetary demise: "But now a cheerful optimist comes forward and tells the race that it has hardly chipped the shell, that the men of today are living in the feeble dawn not so much of civilization as of existence itself. . . . It means that man is 'the heir of all the ages, in the foremost files of time' in a vast and wonderful sense whereof he never has dreamed."[48] Substituting the word "American" for "man" in this passage perhaps gives a better sense for its specific applicability and reception in the United States.

Much has been made of the self-professed "exceptionalism" that Americans have invoked to characterize their cultural and economic history as fundamentally different from those that unfolded in Europe. Instead of colonizing settled lands and forcibly subjugating native peoples, the story goes, early settlers to the American continent focused their efforts on empty lands and otherwise engaged in cooperative relationships with indigenous peoples. Instead of acquiring additional territory through military conquest, the story continues, the young American government used scientific survey and descriptive ethnology to broker the peaceful appropriation of settled Western territories.[49] Instead of engaging in violent European-style imperialism beyond its borders, the story concludes, the emerging American nation pursued a morally superior course of entry into world affairs via peaceful private enterprise. Although much excellent scholarship has pointed out that the deeply ingrained belief in American exceptionalism necessarily glosses over the extensive similarities between American and European forms of expansion, it is nonetheless clear that American expansion had distinct geographical inflections.[50] Smith argues that the late industrialization and economic growth of the United States left the newcomer nation with few territorial options on a globe that had already been fully appropriated by the European nations, save for "the geographical crumbs of an already disintegrating Spanish Empire." It was not so much an aversion to imperialism as a lack of available territory, then, which set the United States on a course toward economic imperialism rather than territorial expansion. Even in this new guise, American economic and geopolitical activities were nonetheless driven by the same basic imperialist mindset that dominated European affairs and indeed the wider western liberal tradition.[51]

As reflected in academic and popular geography, the new American imperialism was less concerned with cultural hierarchy than with the potential for commercial control of natural resources. In the newly acquired Philippines, for example, geographers helped focus intense popular interest in the American imperial effort on its commercial aspects. The moral imperative of confronting, reforming, or even understanding the racialized Filipino Other was accordingly presented to the public as a secondary benefit of commercial activity, rather than a point of focus. Modifying the environmental determinism they had first adopted from European geographers, American geographers thus began to treat cultural difference as the product not of climate and race, but as the product of climate and commerce. As Susan Schulten has argued, "This was a world organized around commercial potential rather than racial difference."[52] Others scholars have argued that the defining element of American imperialism was a focus on the "manipulation of image and spectacle." It was this element, Said has argued, that required

Americans to rely more on mass media and cultural control than on direct dominance of peoples and territories.[53]

Returning to the case of Mars science, the generally enthusiastic American response to the postulation of a superior Martian Other helps illuminate this broad American reframing of imperialism and its theoretical justifications. First and foremost, the American willingness to consider the hypothetical Martian as a nondangerous mentor indicates a revised imaginative geography of the Western encounter with the Other. In his original argument, Said had noted that his concept of Orientalism had limited applicability beyond the European context: "Americans will not feel quite the same about the Orient, which for them is much more likely to be associated very differently with the Far East. . . . To speak of Orientalism therefore is to speak mainly, although not exclusively, of a British and French cultural enterprise." American Orientalism, then, if it can even be properly termed as such, was concerned much less with maintaining "the pattern of relative strength between East and West" than with using science and technology "to define and contain a world in which the American presence was rapidly expanding."[54]

The ways in which American imperialism diverged from European models also points to what may have been a fundamental rethinking of the nature of the Other. Burke's recent argument in *Eyewitnessing* held that "groups confronted with other cultures" can make sense of those groups in one of only two modes: the other group can be stereotyped positively as a reflection of the self or negatively as an inferior Other.[55] The turn-of-the-century American view of the Martian, however, seems to reject this choice. In the elaborate artwork created to illustrate Wells's speculative 1908 article in the American periodical *Cosmopolitan* (fig. 6.7), for example, visual tropes indicate an amalgam of Burke's two categories. The small nuclear family standing on the balcony represents a Western ideal, while the exposure of the female's breasts uses an essentialized visual trope to convey the vulnerability and inferiority of the Other. Likewise, the scene prominently features carpets and minarets that evoke the Eastern world while also foregrounding viewing instruments suggestive of the Western world.

In my reading, the particular way in which Americans embraced the idea of superior Martians indicates an important differentiation between Others admired from a distance versus Others feared at home. Unlike the British Isles, the United States experienced extremely high levels of foreign immigration at the turn of the century, with much of the influx coming from southern and eastern Europe. As these non-Teutonic regions had not

'THERE ARE CERTAIN FEATURES IN WHICH THEY ARE LIKELY TO RESEMBLE US. AND
AS LIKELY AS NOT THEY WILL BE COVERED WITH FEATHERS OR FUR. IT IS
NO LESS REASONABLE TO SUPPOSE, INSTEAD OF A HAND, A
GROUP OF TENTACLES OR PROBOSCIS-LIKE ORGANS"

Figure 6.7. Illustration for a magazine story on Martian inhabitants, 1908. This image, published to accompany an article about Mars in *Cosmopolitan Magazine* (Wells, "Things that Live on Mars," facing p. 334), shows winged Martians in a futuristic-looking space that incorporates visual tropes of both Eastern and Western cultures.

previously contributed significant numbers of immigrants to North America, concerns about social degeneration or excessive dilution of the American racial "stock" reached panic levels as the flow of unfamiliar immigrants increased. Many scholars have explored the ways this demographic phenomenon spurred an explosion of racist sentiment that was expressed through scientific, political, and social movements focused on eugenics, immigration controls, and both educational and urban-environmental reform.[56]

What is not often commented is how this hysterical fear of foreign Others in the midst of the American landscape and social body contrasted with the composed acceptance of foreign Others that remained outside the territorial space of the United States. On American soil, immigrant foreign Others were perceived as inferior, backward, depraved, dangerous, and incapable of either self-representation or political participation. In short, they were cast as classical Others in the European conceptual model. For those Others that remained distant, however, we see much less of a link drawn between Otherness and inferiority. From the Japanese that Lowell so admired to the Filipinos credited with a strong moral potential to the advanced Martians of the red planet, American audiences seem to have accepted a third category that lay somewhere between Self and Other. Differing national context and imperial experience thus helps explain the existence of fundamental differences in popular narratives about superior Martians on either side of the Atlantic.

Conclusions

Attention to the geography of scientific practice, representation, and reception allows for a more nuanced understanding of how and why the scientific and popular narratives of Mars developed as they did around the turn of the century. The legitimacy of astronomers and observatories depended largely on their physical locations and the ways in which those locations were represented to diverse audiences. In claiming legitimacy on the basis of high-altitude locations in remote mountains, furthermore, astronomers found themselves participating in representational conventions and methodological practices drawn from the field sciences. In the process, astronomical Mars science became strongly associated with geographical science, conferring geographical relevance on astronomers' claims about Mars. Because geographical science was conditioned at that time by its engagement with imperialism and expansionism, the debates over Mars must be viewed in light of that particular geopolitical context. The peculiarities of turn-of-the-century Mars science finally begin to make sense from this contextual perspective. Claims previously characterized as illogical emerge as almost self-evident to the wide audiences that accepted them. Practices previously dismissed as unscientific can be redefined as rather savvy moves within the intellectual and social contexts that governed their success. And representations previously ridiculed as overly sensational take on decidedly mainstream dimensions alongside other scientific representations emerging from the field sciences.

The fundamental construction of the Martian landscape in the 1890s as "artificial," or patterned by the activities of intelligent beings, altered long-standing analogies to include Earth's manmade structures as a point of comparison between the two planets. Where the inexplicably geometric appearance of the Martian surface had once defied analogy, Lowell successfully introduced the idea that Mars' physical geography could be equated with Earth's engineered or cultivated landscapes. In advancing this hypothesis, he gained the attention of popular audiences who quickly accepted the idea of an inhabited Mars. At the same time, Lowell managed to overpower most other explanations for Mars' physical appearance, thus constraining the scientific discourse. Lowell's most powerful construction of the Martian landscape painted the planet as a site of tremendous aridity, nourished only by an extensive irrigation system. This representation of Mars as a desert planet relied on frequent and specific comparisons to individual deserts in Africa and Arizona, quickly introducing climatic stereotypes that circulated in much geographical literature at the time. The focus on irrigation, especially, concentrated on a theme that was then a staple of geographic interest in both Europe and North America. Lowell thus presided over a shift in the Martian narrative that saw strangeness converted to familiarity, as the planet's puzzling landscape geometry was said to reveal one of the oldest technologies known to man.

Intimately linked with the discussion of Mars' aridity was the commentary on its continually increasing aridification. Though there was no observational evidence whatsoever to support this claim, Lowell succeeded in painting the red planet as a lost paradise that was suffering the late stages of water loss and desert growth. Writers and audiences responded to this portrayal with very little hesitation, probably because it drew from the standard tropes of desiccation, despoliation, and mismanagement used to represent Earth's arid regions. Although the dominant Lowellian narrative did not hold Martian inhabitants responsible for their planet's imminent demise, it nonetheless exhibited many of the same elements present in geographers' linking of terrestrial landscape with human culture. Following in the environmentally deterministic footsteps of the day's leading geographers, Lowell's assumptions about Martian climate led him to even greater assumptions about the probable intelligence and advancement of the supposed Martian inhabitants. Rather than being seen as dangerous leaps of logic, assertions in this vein were enthusiastically accepted by many of his readers. Similarly, Lowell used visible Martian landscape patterns to support his broad assumptions about Martian civilization, arguing that the complexity of the landscape indicated a certain level of sophistication for the invisible inhabitants.

All of these maneuvers employed standard geographical tropes that built on one another, quickly creating a dominant imaginative geography of Mars. In the process, these tropes also allowed Mars to become a site of projection for terrestrial concerns. Terrors regarding Earth's aridification and dreams about human technological progress, for instance, were expressed and negotiated in arguments and speculations about Mars. As these hopes and fears regarding Earth's geographical change were projected onto Mars, the planet became sensationally popular, thus underscoring the relevance and significance of the Mars narrative well beyond the confines of disciplinary astronomy. The projection of intelligent beings onto Mars notably drew from the Orientalist tradition of representing the Middle East as Europe's polar opposite. Yet some of the most enthusiastic interest in Mars science revolved around narratives that challenged this tradition by casting the Martian civilization as superior to Earth's Western cultures. The trope of the reverse gaze, which posited a Martian observer possessed of the scientific desire and technical ability to observe Earth from afar, was only one of the ways in which the imagined Martians were represented as superior. The assumed facts of Martian evolution were also said to have endowed them with physical, social, intellectual, and technological gifts that far outstripped any known to exist on Earth.

The greater importance of Martian imaginative geographies, however, lies in the ways that different audiences responded to different narratives or claims. By far, the most passionate response to the Lowellian narrative came from American audiences, who took to the deterministic representation of an incredibly superior Martian with very little of the fear that was expressed in other national contexts, such as Britain. This phenomenon can be explained partially by examining the ways that Lowell's specifically American view of the Other was communicated in his claims about Mars. The American interest in commerce and technology as guides to cultural contact, for instance, differed significantly from the European model of racial separation. As Lowell took up themes of cultural contact that he had already considered during travels to the Far East, he made them rhetorically palatable for American audiences in ways that never quite appealed to their European counterparts. The construction of a cultural geography for Mars, steered by the American Orientalism of Lowell and sensationalized by the reaction of his American audiences, thus reframed the cultural encounter with the Other in significant ways.

As Fraser MacDonald recently argued, the exploration of space "from its earliest origins to the present day, has been about familiar terrestrial and ideological struggles here on Earth." If he is correct that the spacefaring activities of the post-Sputnik age have their "origins in older imperial enterprises," I suggest that this condition is only more obvious in the earlier telescopic era that has provided the focus of this book.[57] The geopolitical inflections and implications of the first detailed claims about Martian geography took shape during an era of high imperialism and active expansionism across the Western world, strongly influencing both scientific and nonspecialist understandings of the red planet. In setting the tone for decades of research and popular interest to follow, the sensational debates over Martian canals have not faded into irrelevance as much as they have continually blossomed anew. From science fiction to media coverage to technical investigations of the potential for human settlement on Mars, modern treatments of Earth's orbital neighbor continue to draw heavily on the intellectual currents of turn-of-the-century imperial geography.[58] Lowell's hyperadvanced irrigation engineers may have been given up as ghosts long ago, but the vision of a technologically controlled Martian globe continues to fascinate scientists, policymakers, and generalist audiences. If humans indeed walk on Mars as projected within the next few decades, they will hardly embark on anything "new" but will instead act out a geopolitical vision of human–environment relationships that has been a long century in the making.

NOTES

CHAPTER ONE

1. "Life in Mars," 371.
2. Ibid., 370.
3. Dobbins and Sheehan, "Canals of Mars Revisited."
4. See especially Hoyt, *Lowell and Mars*; Crowe, *Extraterrestrial Life Debate*; Hetherington, "Planetary Fantasies"; and Dick, *Plurality of Worlds*.
5. See especially Markley, *Dying Planet*; Strauss, *Percival Lowell*; Dolan, "Percival Lowell"; Schroeder, "Message from Mars."
6. For more on the intellectual history of the "plurality of worlds" debate see Crowe, *Extraterrestrial Life Debate*; Dick, *Plurality of Worlds*; Guthke, *Last Frontier*; and Markley, *Dying Planet*.
7. Markley, *Dying Planet*. See especially chap. 1.
8. Holden, "What We Really Know," 360.
9. Heward, "Mars," 741.
10. Lockyer, *Elementary Lessons in Astronomy*, 120, 123.
11. Lowell, "Mars: The Water Problem," 750.
12. Schiaparelli, *Astronomical and Physical Observations*, 49, 52.
13. Schiaparelli, "Planet Mars" (no. 129), 714, 719.
14. Holden, "Note on the Mount Hamilton Observations," 668.
15. Ball, "Mars," 203.
16. Gregory, "Mars as a World," 23.
17. Lowell, *Mars*, 2–3.
18. Markley, *Dying Planet*.
19. Mee, *Observational Astronomy*, 55, 52.
20. Crowe, *Extraterrestrial Life Debate*.
21. Edward E. Hale, "Latest News" (*Scientific American*), 137.
22. Ibid.
23. Ibid.
24. Strauss, *Percival Lowell*.
25. Lowell maintained an active and amiable correspondence with W. W. Payne, editor of *Popular Astronomy*, even inviting him to Flagstaff to conduct astronomical observations; Percival Lowell Correspondence, Lowell Observatory Archives, Flagstaff, Arizona (hereafter LOA).

26. See correspondence between Wrexie Louise Leonard (Lowell's personal assistant) and Victor B. Emanuel (a publicist retained by the observatory); Wrexie Louise Leonard Correspondence, LOA.

27. Lowell, *Mars and Its Canals*, viii.

28. Lowell, "On the Portents of Socialism," December 8, 1910; Lowell, "An Address of Welcome to the Good Roads Association," October 8, 1915; Lowell, "Immigration Versus the United States: An Address Delivered at Phoenix, Arizona," February 17, 1916. All of the foregoing are unpublished lectures in Percival Lowell Unpublished MSS, LOA.

29. "The Man Who Made the Martians Live," *Lawrence Evening Tribune*, November 15, 1916; "The Mars Man." *Paterson Guardian*, December 1, 1916.

30. Hilgartner, "Dominant View of Popularization," and Whitley, "Knowledge Producers." See also Secord, *Victorian Sensation*.

31. Hilgartner, "Dominant View," 522. See also Shinn and Whitley, *Expository Science*.

32. Lightman, "Voices of Nature," 187; Henson et al., eds., *Culture and Science*; Cantor and Shuttleworth, eds., *Science Serialized*; and Broks, *Understanding Popular Science*.

33. Flammarion, *La planète Mars*; Proctor, *Other Worlds Than Ours*.

34. Holden, "Latest News," 638.

35. Keeler, "Picturing the Planets," 458.

36. Simon Newcomb to Percival Lowell, September 21, 1905, Lowell Correspondence.

37. Orr, "Canals of Mars," 39.

38. Bailey, "Astronomical Notes," 7.

39. See H. M. Arden to Simon Newcomb, November 25, 1907, Simon Newcomb Papers, Manuscript Division, U.S. Library of Congress (hereafter Newcomb Papers).

40. Dartnell, "A Living Mars?" and Jones, "Mars before the Space Age."

41. For institutional dynamics see Lankford, "Amateurs and Astrophysics"; Lankford, "Amateurs versus Professionals"; Strauss, *Percival Lowell*; Rothenberg, "Organization and Control"; and Hetherington, "Amateur versus Professional." For psychological influences see Sheehan, *Planets and Perception*. For theological and evolutionary debates see Crowe, *Extraterrestrial Life Debate*; Markley, *Dying Planet*; and Dick, *Biological Universe*.

42. See Dick's *Biological Universe* for a biological view of Mars science, and Markley's *Dying Planet* for a tight focus on ecological ideas.

43. Shapin and Schaffer, *Leviathan and the Air-Pump*, 13.

44. Golinski, *Making Natural Knowledge*, ix.

45. Keller, *Reflections on Gender*; Collins, *Changing Order*; Latour and Woolgar, *Laboratory Life*; Latour, *Science in Action*; Collins, "The TEA Set"; Callon, "Elements of a Sociology"; Haraway, "Teddy Bear Patriarchy"; Bassett, "Whatever Happened?" and Pedynowski, "Science(s)—Which, When and Whose?"

46. Foundational works include Shapin and Schaffer, *Leviathan and the Air-Pump*, and Biagioli, *Galileo*. For important examples of recent work, see Prakash, *Another Reason*; Hart, "Translating the Untranslatable"; Galison, "Trading Zone"; Haraway, *Modest_ Witness*; and Latour, *We Have Never Been Modern*.

47. Shapin, "House of Experiment"; Shapin and Schaffer, *Leviathan and the Air-Pump*; Schaffer, "Physics Laboratories"; and Gooday, "Premisses of Premises."

48. Ophir and Shapin, "Place of Knowledge," 4.

49. Livingstone, "Geography, Tradition and the Scientific Revolution," 338; and Livingstone, "Spaces of Knowledge," 16.

50. For overviews, see Finnegan, "Spatial Turn"; Powell, "Geographies of Science"; and Shapin, "Placing the View." For a collection of early work in this vein, see Smith and Agar, *Making Space for Science*. For other recent treatments, see Bourguet, Licoppe, and Sibum, *Instruments, Travel and Science*.

51. Thrift, Driver, and Livingstone, "Geography of Truth," 2.

52. Livingstone, "Spaces of Knowledge"; Livingstone, *Putting Science in Its Place*.

53. Naylor, "Introduction," 6. Anne Secord's oft-cited "Science in the Pub" laid the foundation for much of this work.

54. Finnegan, "Natural History Societies." See also Taylor et al., "Geohistorical Study."

55. Dritsas, "Lake Nyassa to Philadelphia"; Livingstone, "Science, Text and Space"; and Topham, "View from the Industrial Age."

56. Schiaparelli, "Topografia e clima." The draft for Schiaparelli's second major memoir on the planet Mars, in fact, was handwritten on the back of correspondence received from such geographically inclined institutions as the Italian Alpine Club, the Society for Commercial Exploration in Africa, the Third International Geographic Congress, the Society for the Promotion of Scientific Exploration, the Italian Geographical Society, the Geographical Institute, and the Italian Meteorological Association, among others. Schiaparelli, "Marte. Capitolo III. Osservazioni sull'aspetto presentato dalle vari regioni del pianeta durante l'opposizione 1879," l'Archivo Storico dell'Osservatorio Astronomico di Brera, Fondo Schiaparelli.

57. Theodore Roosevelt to Alexander Agassiz, December 26, 1902, Newcomb Papers.

58. Strauss, *Percival Lowell*.

59. For recent reviews of their relevance for geographers, see Withers, "History and Philosophy of Geography"; Naylor, "Historical Geography: Knowledge"; Naylor, "Historical Geography: Geographies and Historiographies"; Barnes, "History and Philosophy"; and Mayhew, "Historical Geography."

60. Driver "Geography's Empire," 28, 23. See Mayhew, "Historical Geography," for a contention that virtually all recent scholarship in historical geography is at least implicitly "working in dialogue with, and in remits developed by, the Foucauldian corpus," 2.

61. Hudson, "New Geography," 12.

62. See chapter 6 for an extensive discussion of the scholarship on geography's early commitment to environmental determinism.

63. Godlewska and Smith, *Geography and Empire*; MacKenzie, *Imperialism and the Natural World*; Bell, Butlin, and Heffernan, *Geography and Imperialism*.

64. Davis, *Resurrecting the Granary of Rome*; Grove, *Green Imperialism*; Grove, *Ecology, Climate and Empire*; and Arnold and Guha, *Nature, Culture, Imperialism*.

65. For recent examples, see Cosgrove and della Dora, *High Places*; Lambert, Martins, and Ogborn, "Historical Geographies of the Sea"; Driver and Martins, eds., *Tropical Visions*; and Kenny, "Colonial Geographies."

CHAPTER TWO

1. Newcomb, "Fallacies about Mars," 11.

2. See, for example, Green, "Notes on the Coming Opposition"; Green, "Approaching Opposition"; Maunder, "Mars Section"; Cammell, "Mars Section."

3. Lightman, "Visual Theology."

4. The argument in this paragraph relies heavily on Lightman's "Visual Theology."

5. Monmonier, *Rhumb Lines and Map Wars*.

6. Edgerton, "From Mental Matrix to *Mappamundi*."
7. Blunck, *Mars and Its Satellites*.
8. Proctor, "Names of Markings on Mars."
9. Although an opposition occurs every twenty-six months when Earth swings past Mars, a perihelic opposition occurs only once about every fifteen years.
10. McKim, "Nathaniel Everett Green."
11. Green, "Observations of Mars," 39.
12. "In Memoriam: Nathaniel E. Green."
13. McKim, "Nathaniel Everett Green."
14. Green, "Observations of Mars"; "Meeting of the Royal Astronomical Society, April 12, 1878"; "Meeting of the Royal Astronomical Society, December 13, 1878."
15. G. V. Schiaparelli, observation logbooks, "Vol. 1 del refrattore di Merz," "Refrattore di Merz, tomo II," l'Archivo Storico dell'Osservatorio Astronomico di Brera.
16. Schiaparelli, *Astronomical and Physical Observations*, 5.
17. Schiaparelli, "Osservazioni astronomiche" (1877–78).
18. Green, "Observations of Mars," 123.
19. Schiaparelli's detailed map included only his own observational data, although his correspondence files show that he certainly communicated with colleagues such as François Terby and Otto Struve.
20. Schiaparelli, *Astronomical and Physical Observations*, 46.
21. Green, "Observations of Mars," 140.
22. "Meeting of the Royal Astronomical Society, April 12, 1878," 122.
23. Nathaniel E. Green to G. V. Schiaparelli, March 15, 1878, in Osservatorio Astronomico di Brera, *Corrispondenza su Marte*, 1:14. G. V. Schiaparelli to Otto Struve, July 6, 1878, in ibid., 1:14–18.
24. Terby, "Nomenclature of Martial Markings," 47 (emphasis in original).
25. Green, "Approaching Opposition," 433.
26. Schiaparelli, "Osservazioni astronomiche, memoria seconda"; Schiaparelli, "Osservazioni astronomiche, memoria terza"; Schiaparelli, "osservazioni astronomiche, memoria quarta"; "The Canals on Mars," *Astronomical Register*; Perrotin, "Observation des canaux."
27. Biographical and professional details about Schiaparelli are contained in the book presented to him upon his retirement from the Brera Astronomical Observatory: *All'Astronomo G. V. Schiaparelli*.
28. "Meeting of the Royal Astronomical Society, April 14, 1882"; "Schiaparelli's Observations of Mars."
29. Green, "Observations of Mars," 130. For a full discussion of Green's identity as a professional artist, see McKim, "Nathaniel Everett Green."
30. "Report on the Meeting of the Association held December 31, 1890," 110–11.
31. "Meeting of the Royal Astronomical Society, April 12, 1878," 123.
32. For influential works see Harley, "Deconstructing the Map"; Wood and Fels, "Designs on Signs"; Wood and Fels, *Natures of Maps*; Edgerton, "From Mental Matrix to *Mappamundi*"; Boelhower, "Inventing America"; Harley, "Maps, Knowledge, and Power"; Edney, *Mapping an Empire*; Ryan, *The Cartographic Eye*; Cosgrove, *Mappings*; Cosgrove, *Apollo's Eye*.
33. Schiaparelli, *Astronomical and Physical Observations*, 40.
34. Noble, "Names of Markings on Mars," 96; Marth, "Nomenclature of Markings," 24.
35. Schiaparelli, *Astronomical and Physical Observations*, 10 (italics in original).
36. Ibid., 9.

37. Schiaparelli, observation logbook, "Refrattore di Merz, tomo II."
38. Fennessy, "Nomenclature of Markings," 90 (italics in original).
39. Gill, "Nomenclature of Markings," 95 (italics in original).
40. Noble, "Names of Markings on Mars," 96.
41. Schiaparelli, *Astronomical and Physical Observations*, 9–10.
42. "Past Opposition of Mars," 286.
43. Holmes, "Canals of Mars," 302.
44. Burton, "Canals on Mars," 142; Proctor, "Note on Mars"; Pickering, "An Explanation"; Pickering, "Planet Mars."
45. The major works by William H. Pickering and Alfred Russel Wallace provide good examples of this perspective.
46. E. Walter Maunder, Eugène M. Antoniadi, and Simon Newcomb fall into this category.
47. Schiaparelli, "The Planet Mars," no.129, 719, 722.
48. Driver, "Exploration by Warfare."
49. For a set of now-classic works exploring the unique authority of the map as a visual text, see Harley, *New Nature of Maps*.
50. Bassett and Porter, "From the Best Authorities."
51. Schiaparelli's file on printer proofs for his Mars graphics and maps includes a plate of Green's 1877 sketches, marked with a handwritten note to his publisher: "Queste Tavola si danno per indicare la qualitá del rosso che si deve adoperare e lo stile del lavoro" ("This plate is sent to indicate the quality of red that should be adopted and the style of the work"). "Disegni e 'Mappe Areographiche' di Marte," l'Archivo Storico dell'Osservatorio Astronomico di Brera, Fondo Schiaparelli.
52. Hoyt, *Lowell and Mars.*
53. Simon Newcomb to Percival Lowell, October 30, 1905, Newcomb Papers.
54. Payne, "The 'Canals' of Mars," 366.
55. "French Clergyman Combats Theory."
56. Agassiz, "Mars As Seen in the Lowell Refractor," 281.
57. Antoniadi, "Section for the Observation of Mars: 1909," 31 (italics in original).
58. See, for example, Green, "Northern Hemisphere of Mars"; Maunder, "Section for the Observation of Mars: 1892"; Antoniadi, "Section for the Observation of Mars: 1896."
59. Lowell described his cartographic process in his first book, *Mars.*
60. See, for example, Holmes, "Notes Re Mars."
61. Lowell, "Mars," *Popular Astronomy* (1894): 8; Lowell, "Mars: Oases," *Popular Astronomy* (1895): 346 (emphasis added); Pickering, "The Planet Mars," 469–70.
62. Hoyt, *Lowell and Mars*; Strauss, *Percival Lowell.*
63. "Report of the Meeting of the Association, June 20, 1906," 333; Holmes, "Communication with Mars," 202; "Planet Mars," *Popular Astronomy* (1907): 449–50; "'Signals from Mars," 47.
64. "Report of the Meeting of the Association, Held on June 24, 1903," 338.
65. Strauss, *Percival Lowell.*
66. See, for example, two letters Simon Newcomb wrote to Percival Lowell, March 9 and 23, 1903, Lowell Correspondence.
67. See Galison, "Judgment Against Objectivity."
68. "Canals on Mars," *Scientific American Supplement*, 25143; Kaempffert, "What We Know about Mars," 482; "Report of the Meeting of the Association, June 20, 1906," 333.
69. Newcomb, "Optical and Psychological Principles."

70. See especially Lowell's personal correspondence with Simon Newcomb and Walter Maunder, e.g., Percival Lowell to E. W. Maunder, November 28, 1903; Percival Lowell to Simon Newcomb, January 5, 1907; and Percival Lowell to Simon Newcomb, May 15, 1907, in Lowell Correspondence.

71. Tucker, "Photographic Evidence."

72. R. U. Johnson to Percival Lowell, September 24, 1907; George R. Agassiz to Percival Lowell, September 27 and October 14, 1907; R. U. Johnson to Percival Lowell, October 8, 1907: all in Lowell Correspondence.

73. Tucker, "Photographic Evidence."

74. Crommelin, "Martian Photography"; Antoniadi, "Mars Section Interim Report on the Australian Observations, 1907," 401.

75. Lowell, "New Photographs of Mars," 309–10 (emphasis added).

76. Hugh Chisholm to Simon Newcomb, February 5, 1907, Newcomb Papers (emphases in original).

77. To clarify (or perhaps complicate) the description of Antoniadi as "French": Antoniadi was an ethnically Turkish, Greek-born, naturalized French citizen who became active in the British astronomical community during residence in London. For a detailed discussion of Antoniadi's long involvement in the Mars debate, see McKim, "Antoniadi, Part 2."

78. Antoniadi, "Mars Section Third Interim Report for 1909," 28.

79. E. M. Antoniadi to Percival Lowell, October 9, 1909, Lowell Correspondence.

80. Percival Lowell to E. M. Antoniadi, September 26, 1909, and Antoniadi to Lowell, October 9, 1909 (emphasis in original), Lowell Correspondence.

81. Ibid.

82. Percival Lowell to E. M. Antoniadi, November 2, 1909, and Antoniadi to Lowell, November 15, 1909, Lowell Correspondence.

83. Lowell, "Schiaparelli."

84. Antoniadi, "On the Possibility of Explaining," 93.

85. Antoniadi, "Mars Section Fourth Interim Report for the Apparition of 1909," 79; Antoniadi, "Mars Section Fifth Interim Report for 1909," 141 (emphasis in original).

86. "Report of the Meeting of the Association, Held on Wednesday, December 29, 1909."

87. See, for example, Antoniadi, "Mars Section, Second Interim Report for 1898–99"; Antoniadi, "Section for the Observation of Mars: 1900–1901"; McKim, "Antoniadi, Part 1" and "Antoniadi, Part 2."

88. Antoniadi, "Note on Some Photographic Images," 110, 112; Antoniadi, "Mars Section Sixth Interim Report for 1909."

89. Antoniadi, "Mars Section Fifth Interim Report for 1909," 141.

90. Hoffman, "The Canals of Mars," 362.

91. "Report of the Meeting of the Association, Held on Wednesday, December 29, 1909," 123.

92. Markley's Dying Planet cites mid-century quotes that referred to Lowellian conclusions as having been "refused general acceptance" (163) during a period in which "professional astronomers did not agree with Lowell" (165). Although canalist sentiment certainly persisted and allowed for a mid-century revival of canal-related debates, even the canalists admitted that general scientific consensus had largely rejected Lowell's hypothesis.

93. Edney, Mapping an Empire; Gregory, "Between the Book and the Lamp"; Ryan, Cartographic Eye.

94. Markley, *Dying Planet*.
95. Lowell, "Mars: The Flagstaff Photographs," 643.

CHAPTER THREE
1. Clerke, *Popular History of Astronomy*.
2. Lankford, *American Astronomy*.
3. "Schiaparelli's Observations of Mars," 143.
4. Douglass, "Scales of Seeing," 16.
5. Cited in Holden, *Mountain Observatories*, 1.
6. Ibid., iii.
7. Ibid., 3 (emphasis in original).
8. Ibid., 10.
9. Osterbruck, Gustafson, and Unruh, *Eye on the Sky*.
10. Pickering, "Mountain Observatories," 100.
11. Osterbruck, Gustafson, and Unruh, *Eye on the Sky*.
12. Holden, "Lick Observatory," *Sidereal Messenger* (1888).
13. "Lick Observatory," *Sidereal Messenger* (1885), 49.
14. Holden, *Brief Account of the Lick Observatory*.
15. Osterbruck, *Yerkes Observatory*.
16. Ibid., 15, 16.
17. Holden, *Mountain Observatories*, 50.
18. Hale, "Yerkes Observatory."
19. Hale, "Aim of the Yerkes Observatory," 310.
20. Hale, "Yerkes Observatory," 168, 177.
21. Hale, "Aim of the Yerkes Observatory," 317.
22. Cited in Osterbruck, *Yerkes Observatory*, 30.
23. First announced in Hale, "Development of a New Observatory."
24. Lowell, *Mars*, v.
25. Lowell, "New Photographs of Mars," 303.
26. Newcomb, "Astronomy," 728; Lowell, "On the Climatic Causes"; *The Observatory* 19, no. 239 (1896): 177–78.
27. Flammarion, "Recent Observations of Mars," 133.
28. Douglass, "Lowell Observatory and Its Work," 395.
29. Strauss, "Percival Lowell, W.H. Pickering"; Fay et al., *Century of American Alpinism*.
30. Schama, *Landscape and Memory*.
31. Macfarlane, *Mountains of the Mind*, 159–60.
32. Holden, "Lick Observatory," *New York Daily Tribune*.
33. Holden, "What We Really Know."
34. Campbell, "Spectrum of Mars," *Publications of the Astronomical Society of the Pacific*, 230.
35. "Schiaparelli's Observations of Mars." 143.
36. P. Lowell to W. Maunder, November 28, 1903, Lowell Correspondence.
37. "Lick Observatory at the University of California," 162.
38. Lowell, *Mars and Its Canals*, 16, 17.
39. Ibid., 18.
40. "Report of the Meeting of the Association, Held on March 30, 1910," 290.
41. Lowell, *Mars and Its Canals*, 7, 8.
42. Pickering, "Planet Mars," 463–64.
43. The following characterization is drawn largely from Nash, *Wilderness and the American Mind*.

44. Lowell, *Mars and Its Canals*, 15.
45. "Life at the Lick Observatory," 73.
46. Ibid.
47. See Perrine, "List of Earthquakes in California."
48. "Life at the Lick Observatory," 73.
49. Ibid.
50. Lowell, *Mars and Its Canals*, 6.
51. Lowell, "New Photographs of Mars," 303.
52. Douglass, "Mars," 113.
53. Percival Lowell to David Gill, January 23, 1913, Lowell Correspondence.
54. Morse, *Mars and Its Mystery*.
55. Agassiz, "Mars as Seen in the Lowell Refractor," 275.
56. Hale, "Latest News from Mars," *Publications of the Astronomical Society of the Pacific*, 117.
57. Campbell, "Spectrum of Mars as Observed," 161.
58. Kohler, *Landscapes and Labscapes*. But I am indebted to Pilkington's review of Kohler ("Ecologist's Very Own Ecotone") for noting that his neat distinction between lab and field is perhaps insufficient in contemporary contexts, as a more relevant distinction has emerged between observational and experimental practice.
59. Driver, "Making Space."
60. See also Gieryn, *Cultural Boundaries*; Kuklick and Kohler, "Science in the Field"; and McCook, "It May Be Truth."
61. Schaffer, "Astronomers Mark Time."
62. Driver, "Making Space," 389.
63. Simon Newcomb to Percival Lowell, May 8, 1905, Lowell Correspondence.
64. The following summary is drawn from Lankford, "Amateurs versus Professionals"; Lankford, "Amateurs and Astrophysics"; and Rothenberg, "Organization and Control."

CHAPTER FOUR
1. Schiaparelli, *Astronomical and Physical Observations*, 3, 1.
2. This term had been used as early as 1868 by British science writer Richard. A. Proctor (see Lightman, "Visual Theology"), but it became widespread after Schiaparelli's 1878 publication.
3. Lowell, *Mars*, 93.
4. There are numerous instances of this phenomenon, particularly in *Scientific American*.
5. E.g., Newcomb, "Mars."
6. Lowell, "Mars: The Polar Snows," 54; Lowell, "Mars: The Water Problem," 750; Lowell, "New Canals of Mars," 190.
7. David, *Arctic in the British Imagination*, 2. See also Bloom, *Gender on Ice*; Hill, *White Horizon*; Berton, *Arctic Grail*.
8. Gregory, "Mars as a World," 22.
9. Brewster, "Earth and the Heavens," 262.
10. Rose, "Geography as the Science of Observation."
11. Lutz and Collins, *Reading National Geographic*; Driver, "Geography, Empire and Visualisation"; Ryan, *Cartographic Eye*; Schwartz, "*Geography Lesson*"; Cameron, *Looking for America*.
12. Rose, "Geography as the Science of Observation," 9.
13. Lowell, *Mars*, 93; Lowell, "Mars: The Polar Snows," 55; Flammarion, "Mars and Its Inhabitants," 546; "Publications: Mars," 281.
14. Lowell, *Mars*, 112.

15. Lowell, "Mars: The Canals," 257–58.
16. Lowell, "New Photographs of Mars," 310.
17. Morse, *Mars and Its Mystery*, vii–viii.
18. Scientists now believe the white patches at Mars' poles indeed contain water ice, but are largely deposits of carbon dioxide, which sublimates (without melting) at −109°F, well below the warm temperatures that would be required to melt a similar extent of water ice.
19. Manson, "Climate of Mars." The idea that Mars' temperature was comparable to Earth's was presented as common knowledge in general publications such as *Reynold's Universal Atlas*; Chambers, *Handbook of Descriptive Astronomy*; and Newcomb and Holden, *Astronomy for Schools*.
20. Schiaparelli, "Planet Mars," 640. Note: This was the second of two installments that appeared under the same title.
21. Lepper, "Examination of the Modern Views"; Holt, "Solar Image Reflected"; "Image of the Sun"; Pickering, "Mars" and "Colors Exhibited" nos. 106 and 107.
22. "Atmosphere of Mars." See also the discussion of Campbell's 1909 spectroscopic work later in this chapter.
23. Ball, *In the High Heavens*, 145.
24. Rose, "Geography as the Science of Observation," 9. See also Driver, "Editorial: Field-Work in Geography"; Withers, "History and Philosophy."
25. See, e.g., Newcomb, *His Wisdom the Defender*; Schiaparelli, "Gli Abitanti di Altri Mondi."
26. Lowell, *Mars*, 92–93, 94.
27. For more detail than the following discussion provides, see Jones and Boyd, *Harvard College Observatory*; Plotkin, "Harvard College Observatory's Boyden Station"; Bailey, "Expeditions and Foreign Stations"; and Nisbett, "Business Practice."
28. Jones and Boyd, *Harvard College Observatory*.
29. Lankford, *American Astronomy*; Nisbett, "Business Practice."
30. Bailey, "History of the Expedition." See, e.g., Fleming, "Harvard College Observatory Astronomical Expedition"; and "Mapping the Southern Sky."
31. Fleming, "Harvard College Observatory Astronomical Expedition," 59; Todd, "Great Modern Observatory," 297. For more on this, see Bailey, "History of the Expedition," and Bailey, *History and Work of the Harvard Observatory*.
32. Bailey, "History of the Expedition," 31.
33. Holden, *Mountain Observatories*, 35.
34. Pickering, "Climb in the Cordillera," 206.
35. Nisbett, "Business Practice."
36. Pickering, cited in Holden, *Mountain Observatories*, 37.
37. Jones and Boyd, "Some Lofty Mountain," 315. See also "Harvard Observatory in Peru"; Bailey, "Harvard Observatory in Peru," 329.
38. Jones and Boyd, "Some Lofty Mountain," 316.
39. Holden, *Mountain Observatories*, 37, 14.
40. Jones and Boyd, "Some Lofty Mountain."
41. Pickering, "Mars," 675.
42. Nisbett, "Business Practice."
43. Gregory, "Between the Book and the Lamp"; Pratt, *Imperial Eyes*; Kearns, "Imperial Subject"; Morin, "British Women Travellers"; Blunt, "Imperial Geographies of Home"; McEwan, *Gender, Geography and Empire*; Guelke and Morin, "Gender, Nature, Empire"; and Brown, "Richard Vowell's Not-So-Imperial Eyes."

44. Lowell, "New Photographs of Mars," 304.
45. For a complete account of this expedition, see Todd, "Lowell Expedition to the Andes" and "Professor Todd's Own Story."
46. Todd, "Professor Todd's Own Story," 349.
47. Lowell, "New Photographs of Mars," 303.
48. Pang, "Social Event of the Season," and "Gender, Culture, and Astrophysical Fieldwork."
49. Percival Lowell to David Todd, July 26, 1907, Lowell Correspondence.
50. Todd, "Photographing the 'Canals,'" and "Our Ruddy Neighbor Planet"; Todd, "Professor Todd's Own Story" and "Lowell Expedition to the Andes"; Slipher, "Photographing Mars"; Lowell, "New Photographs of Mars."
51. S. S. Chamberlain to Percival Lowell, September 6 and 21, 1907; R. U. Johnson to Percival Lowell, September 19, 1907; S. S. Chamberlain to Percival Lowell, October 11, 1907; Johnson to Lowell, September 19, 1907: all in Lowell Correspondence.
52. S. S. Chamberlain to Percival Lowell, October 22, 1907, Lowell Correspondence.
53. R. U. Johnson to Percival Lowell, October 3, 1907, Lowell Correspondence.
54. R. U. Johnson to Percival Lowell, October 5, 1907, Lowell Correspondence.
55. Schulten, *Geographical Imagination*; Smith, *American Empire*.
56. Driver, "Distance and Disturbance," 77.
57. Pratt, *Imperial Eyes*; Brady, "Full of Empty." But see an important critique of this in Guelke and Guelke, "Imperial Eyes on South Africa."
58. In addition to the works cited in n. 43 in this chapter, see also Duncan and Gregory, *Writes of Passage*, and Ryan, "Our Home on the Ocean."
59. Todd, "Professor Todd's Own Story," 348.
60. Todd, "Photographing the 'Canals,'" refers to this as an "English" company, while several American newspapers referred to it as "American."
61. Ibid., 265.
62. Pang, "Social Event of the Season." Fan's conception of "informal empire" as partly secured by scientists through their "conviction of a right to know, a right that was ideally not restrained by human boundaries" also applies here to some extent. Fan, "Victorian Naturalists in China," 25.
63. Todd, "Professor Todd's Own Story," 347.
64. "To Photograph Mars," 1; "Scientists Will Try," 1.
65. Todd, "Professor Todd's Own Story," 348.
66. "'Something Like Human Intelligence on Mars.'"
67. Pang, "Social Event of the Season," 268.
68. Todd, "Professor Todd's Own Story," 350.
69. Ibid., 349, 351.
70. Blunt, "Imperial Geographies of Home." See also Kearns, "Imperial Subject."
71. Driver, "Henry Morton Stanley and His Critics"; Morin, "Charles P. Daly's Gendered Geography."
72. W. W. Campbell to G. E. Hale, May 11, 1908, cited in Sheehan, *Planet Mars*, 131–32.
73. Driver, "Henry Morton Stanley and His Critics," 148.
74. Kearns, "Imperial Subject."
75. This development paralleled the negotiation of boundaries between fieldwork and labwork. See discussion on this topic in chapter 3.
76. Pang, "Gender, Culture, and Astrophysical Fieldwork."
77. Blunt, *Travel, Gender, and Imperialism*; Morin, "Trains Through the Plains"; Gregory, "Between the Book and the Lamp"; Morin, "Peak Practices"; Mills, *Discourses of Difference*.

78. Campbell, "Spectrum of Mars as Observed"; DeVorkin, "W. W. Campbell's Spectro-scopic Study"; Osterbruck, "To Climb the Highest Mountain."
79. Osterbruck, "To Climb the Highest Mountain."
80. DeVorkin, "W. W. Campbell's Spectrographic Study."
81. Lowell, "Spectra of the Major Planets."
82. Campbell, "Spectrum of Mars as Observed," 149.
83. Ibid., 152.
84. Ibid., 152, 155.
85. Ibid., 153, 155.
86. DeVorkin, "W. W. Campbell's Spectrographic Study."
87. Osterbruck, Gustafson, and Unruh, *Eye on the Sky*; Pang, "Gender, Culture, and As-trophysical Fieldwork."
88. Osterbruck, "To Climb the Highest Mountain," 80, 81.
89. Campbell, "Spectrum of Mars as Observed," 152.
90. W. W. Campbell to G. F. Marsh, June 11, 1908, cited in Osterbruck, "To Climb the Highest Mountain," 83 n. 21.
91. Osterbruck, "To Climb the Highest Mountain," 83.
92. Aitken, "William Wallace Campbell."
93. Dritsas, "From Lake Nyassa to Philadelphia"; Hevly, "Heroic Science"; Tucker, "Voy-ages of Discovery"; Morin, Longhurst, and Johnson, "(Troubling) Spaces"; Macfarlane, *Mountains of the Mind*.
94. Halford Mackinder characterized geography and geographical education as the de-velopment of a "rationalizing eye [that] would sweep over gradually widening areas, until at last it embraced the entire surface of the world," giving citizens "the power of roaming at ease imaginatively over the vast surface of the globe"; quoted in Driver, "Geography, Empire and Visualisation."
95. Rose, "Geography as a Science of Observation."
96. Outram, "New Spaces in Natural History."
97. Cosgrove and Daniels, eds., *Iconography of Landscape*.

CHAPTER FIVE

1. Warner, "Editor's Study," 636.
2. Ibid., 636.
3. Ibid., 638, 639.
4. Rose, "Geography as a Science of Observation," 11. Cosgrove and Daniels, *Iconog-raphy of Landscape*, suggests that landscape is better understood as a concept or a "cultural image." See also Cosgrove, *Social Formation and Symbolic Landscape*.
5. The extreme negative view is apparent in Heffernan, "Singularity of Our Inhabited World," 82; the opposite extreme dominates Dolan, "Percival Lowell."
6. "Man Who Explored Mars," unlabeled newspaper clipping, December 2, 1916, Newspaper Clipping Files, 1894–1916, LOA; Markley, *Dying Planet*, xx; Dolan, "Per-cival Lowell."
7. Lowell, *Mars and Its Canals*, viii–ix; Holden, "Lowell Observatory in Arizona," 160.
8. Fichman, *Elusive Victorian*, 211, 212; Shermer, *In Darwin's Shadow*.
9. Alfred Russel Wallace to Macmillan and Co., August 9, 1907, Macmillan and Co. MSS, British Library.
10. See letters from Wallace to Macmillan and Co. on August 15 and September 25, 1907 (Macmillan and Co. MSS) regarding his delays in submitting the manuscript while he waited to consult with "friends."

11. Wallace, *Is Mars Habitable?*, 9.

12. For positive views see Heffernan, "Singularity of Our Inhabited World"; Sheehan, *Planets and Perception*; Sheehan, *Planet Mars*. For critical views see especially Dick, *Biological Universe*.

13. Lowell, "The Geography of Mars: Lecture to the National Geographic Society, Washington, D.C.," January 3, 1908, handwritten notes, Percival Lowell Unpublished MSS, LOA; Wallace, *Is Mars Habitable?*, 20.

14. Flammarion, *La planète Mars*; "Planet Mars," *Observatory*.

15. Lowell, "Mars: Seasonal Changes," 821. Pickering was first to argue that Mars' dark spots were due to vegetation. Pickering, "Seas of Mars."

16. Lowell, "Cartouches of the Canals."

17. Lowell, "Mars: The Water Problem."

18. It seems that Lowell used the term "gravity" to refer to the coriolis effect of a sphere's rotation on the flow of its surface waters, as seen in this typical discussion: "No sooner liberated from its winter fetters than it [water] would begin under the pull of gravity to run toward the equator. (It may interest the reader to note that its course would on the spheroidal surface actually be uphill.) Each particle would start due north; but its course would not continue in that direction. For at each mile traveled north it would reach a latitude of greater rotation than the last it left. . . . The consequence upon the particle would be its northerly motion would be continuously changing with regard to the surface into a more and more westerly one." Lowell, "Mars: Spring Phenomena," 98.

19. Lowell, "The Geography of Mars: Lecture to the National Geographic Society, Washington, D.C.," January 3, 1908, handwritten notes, Percival Lowell Unpublished MSS, LOA.

20. Lowell, "Planet Mars," 224.

21. Lowell, *Mars and Its Canals*, 149.

22. "Will the New Year Solve the Riddle of Mars?"

23. Percival Lowell to Elizabeth Lowell Putnam, February 13, 1896, Letters to Elizabeth Lowell Putnam, 1876–1916, Harvard University Houghton Library.

24. Manson, "Climate of Mars"; Campbell, "Concerning an Atmosphere on Mars."

25. See especially *Is Mars Habitable?*, 57, 106–7, 109.

26. See also Lowell, "Plateau of the San Francisco Peaks."

27. Lowell, *Mars as the Abode of Life*, 97–98.

28. Manson, "Climate of Mars," 374.

29. Markley, *Dying Planet*.

30. Lowell, *Mars as the Abode of Life*, 124, 131, and see especially chapter 4.

31. Lowell, "Mars: Oases," *Atlantic Monthly*, 234. The nebular hypothesis was powerfully challenged in 1905 by the planetesimal (Chamberlin-Moulton) hypothesis, which rejected both the cooling/drying phases of the nebular hypothesis, as well as its provision for common life forms throughout the universe. See Strauss, *Percival Lowell*, for a detailed discussion of Lowell's beliefs on cosmic evolution.

32. Kennedy, "Inventing the Geographical Cycle"; Chorley, "Re-evaluation of the Geomorphic System," 30; Livingstone, "Geographical Experiment."

33. W. M. Davis's "geographical cycle"—or "the Davisian cycle"—is not really a cycle at all, but is rather a unidirectional sequence, which is why Davis found it useful to compare landscape stages to the stages of human life. Kennedy, "Inventing the Geographical Cycle."

34. *Lowell, Mars and Its Canals*, 16.

35. Lowell, *Mars as the Abode of Life*, 119.
36. Huntington's early work on climate was summarized in *Civilization and Climate*.
37. Lowell, *Mars as the Abode of Life*, 122.
38. Lowell, *Mars and its Canals*, 152.
39. Lowell, "The Newly Discovered Petrified Forest of Arizona," Percival Lowell Working Papers, LOA.
40. Blackwelder, "Mars as the Abode of Life," 659, 660, 661.
41. Davis, *Resurrecting the Granary of Rome*. See also Davis, "Potential Forests"; Davis, "Environmentalism as Social Control?"; Grove and Rackham, *Nature of Mediterranean Europe*; Grove, "Evolution of the Colonial Discourse."
42. Davis, *Resurrecting the Granary of Rome*; Grove, "Imperialism and the Discourse of Desiccation." See also Gregory, "(Post)Colonialism and the Production of Nature."
43. Lowell, *Mars as the Abode of Life*, 128–29.
44. Lowell, *Mars and Its Canals*, 153.
45. Ibid., 124.
46. Ibid., 153.
47. Lowell, *Mars as the Abode of Life*, 134.
48. The originator of the theory, T.C. Chamberlin, was head of the geology department at the University of Chicago. Brush, "Geologist Among Astronomers."
49. Markley, *Dying Planet*, 67.
50. Ibid., 68, 69.
51. Lowell, *Mars as the Abode of Life*, 134, 124.
52. Lowell, "Mars: Oases," *Popular Astronomy*, 344; Lowell, "Mars: Canals," 111, 112.
53. Lowell, "Mars: Oases," *Popular Astronomy*, 348.
54. Lowell, "Planet Mars," 224.
55. Strauss, "Fireflies Flashing in Unison," 160.
56. Lowell, "Mars: Canals," 107.
57. Morse, *Mars and Its Mystery*, 118.
58. Markley, *Dying Planet*.
59. "Science," 317.
60. Lowell, *Mars and Its Canals*, 362–63.
61. Hudson "New Geography."
62. Livingstone, "Geographical Experiment," 177; Peet, "Social Origins of Environmental Determinism," 313.
63. Ratzel, *History of Mankind*, 1:27.
64. Livingstone, *Geographical Tradition*, 210.
65. Livingstone, "Sternly Practical Pursuit," 221. See also Livingstone, "Climate's Moral Economy."
66. Driver and Martins, *Tropical Visions*.
67. Semple, *American History*, 77; Turner, "Significance of the Frontier"; Livingstone, "Environment and Inheritance"; Coleman, "Science and Symbol."
68. Lowell, *Soul of the Far East*, 223–24, 225–26. See also Lowell, *Chosön* and *Noto*.
69. Strauss, *Percival Lowell*.
70. Lowell, "Oration on the Fourth of July at Flagstaff, Arizona," July 4, 1901, typed lecture with handwritten notes, Percival Lowell Unpublished MSS, LOA.
71. Semple, *American History*, 231.
72. Livingstone, "Geographical Experiment"; Stoddart, "Darwin's Impact on Geography"; Peet, "Social Origins of Environmental Determinism," 310; Stromquist, *Re-Inventing "The People."*

73. See especially Percival Lowell, "Immigration Versus the United States: an Address Delivered at Phoenix, Arizona," February 17, 1916, and "On the Portents of Socialism," December 8, 1910, typed lecture notes, Percival Lowell Unpublished MSS, LOA. See also Dolan, "Percival Lowell," on the subject of Lowell's political alignment with Roosevelt and the Progressive Party on these issues.

74. Strauss, *Percival Lowell*, 109.

75. Livingstone, *Geographical Tradition*; Stromquist, *Re-Inventing "The People"*; Ward, "Mars and Its Lesson," 163, 165.

76. Wallace, *Is Mars Habitable?*, 104, 105.

77. See especially Markley, *Dying Planet*; Dick, *Biological Universe*.

78. Fichman, *Elusive Victorian*, 260, 261.

79. Ibid., 214, 228, 255.

80. Ibid., 262.

81. Ibid., 277.

82. Hoyt, *Lowell and Mars*, 288.

83. Markley, *Dying Planet*; Wallace, *Man's Place*, 329.

84. Lowell, *Mars as the Abode of Life*, 122, 124.

85. Ward, "Mars and Its Lesson," 164.

86. Marsh, *Man and Nature*.

87. Thomas, *Man's Role*. But see Diamond's best-seller *Collapse* for a new incarnation of environmental determinism.

88. See especially Wittfogel, "Hydraulic Civilizations."

89. Nash, *Wilderness and the American Mind*.

90. Worster, *Rivers of Empire*, 111, 113, 114.

91. Westcoat, "Wittfogel East and West."

92. Powell, *Report on the Lands*; Kirsch, "John Wesley Powell"; Goetzmann, *Exploration and Empire*; Worster, *River Running West*; Smythe, *Conquest of Arid America*; Lee, "William Ellsworth Smythe."

93. Worster, *Rivers of Empire*, 116.

94. Kaempffert, "What We Know about Mars," 481.

95. Morse, *Mars and Its Mystery*, 115, 123.

96. Gregory, "Mars as a World," 24.

97. Lowell, *Mars*, 129.

98. Lowell, *Mars and Its Canals*, 376–77.

99. Ibid., 377–78.

100. Lowell, "Thoth and the Amenthes," 43; Lowell, *Mars and Its Canals*.

101. Lowell, *Mars as the Abode of Life*, 143.

102. Hale, "Latest News from Mars," *Scientific American*, 137. Note: this article was reprinted from *Boston Commonwealth*.

103. "Why the Dwellers on Mars," 214.

104. Wallace, *Is Mars Habitable?*, 32.

105. Ibid., 9.

106. Godlewska, "Map, Text and Image"; Willcocks, *Sixty Years in the East*. See also Heffernan, "Bringing the Desert to Bloom."

107. Lockyer, "Opposition of Mars," 448.

108. Endfield and Nash, "Missionaries and Morals"; Deakin, *Irrigated India*, 20, 234; Willcocks, *Sixty Years in the East*; Hollings, *Life of Sir Colin Scott-Moncrieff*.

109. Wallace, *Wonderful Century*, 376 and especially chapter 21, "The Plunder of the Earth—Conclusions." See also Davis, *Late Victorian Holocausts*.

110. Wallace, *Wonderful Century*, 377.
111. Wallace, *Is Mars Habitable?*, 104, italics in original.
112. Clerke, "New Views about Mars," 379; Wallace, *Is Mars Habitable?*, 103.
113. Ibid.; Housden, *Riddle of Mars the Planet*; "Riddle of Mars"; Lowell, *Mars as the Abode of Life*, 208.
114. Wittfogel, *Oriental Despotism*. Wittfogel's thesis has now been so thoroughly critiqued that it is rarely taken seriously in studies of arid-lands societies and communities. (But see Worster, *Rivers of Empire*, for a spirited defense of its broad applicability.) When originally presented, however, it was seen as an important alternative to the strict environmental determinism that had dominated discourse to that point.

CHAPTER SIX

1. Warner, "Editor's Study," 638, 639.
2. Hall, "West and the Rest"; Hall and Gieben, *Formations of Modernity*.
3. Said, *Orientalism*.
4. For critiques see especially Driver, "Geography's Empire," and Kennedy, "Imperial History." For key works that expand on Said's ideas see Bishop, *Myth of Shangri-La*; Brantlinger, "Victorians and Africans"; Lowe, *Critical Terrains*. For key recent works in geography that respond to Said, see Gregory, "Between the Book and the Lamp"; Kearns, "Imperial Subject"; King, "(Post)Colonial Geographies"; Schwartz, "Geography Lesson"; Lutz and Collins, *Reading National Geographic*; Winlow, "Mapping Moral Geographies"; Harley, "Deconstructing the Map"; Ryan, *Cartographic Eye*; Edney, *Mapping an Empire*.
5. Said, *Orientalism*, 71.
6. See Godlewska, "Map, Text and Image"; Saberwal, "Science and the Desiccationist Discourse"; Grove, "Evolution of the Colonial Discourse"; Grove and Rackham, *Nature of Mediterranean Europe*; Blaut, *Colonizer's Model of the World*; and Davis, *Resurrecting the Granary of Rome*.
7. "Mars and His Moons," 265.
8. Kaempffert, "What We Know about Mars," 481.
9. Lowell, *Mars*, 58. See also Lowell, "Mars: Atmosphere."
10. Montgomery, "Mars," 611; Lowell, *Mars*, 211; Wells, "Things That Live on Mars," 342.
11. Kaempffert, "What We Know about Mars," 486.
12. Morse, "My 34 Nights on Mars," 9.
13. Warner, "Editor's Study," 639.
14. Holmes, "The Canals of Mars," 301–2.
15. Herschel, *Outlines of Astronomy*, 338–39.
16. Wilson, "Mars and His Canals," 14.
17. Unlabeled newspaper clipping, 1905, Newspaper Clipping Files, 1894–1916, LOA.
18. "Notes," *Observatory*, 221.
19. Ryan, "Inscribing the Emptiness."
20. Brewster, "Earth and the Heavens," 262.
21. Wells, "Things that Live on Mars," 342.
22. Sinnott, "Mars Mania."
23. "Notes: The Opposition of Mars," 477.
24. Lowell, "Explanation of the Supposed Signals."
25. Tesla, "Talking with Planets," 4–5.
26. Campbell, "Projections of the Planet Mars," 1903.

27. Flammarion, "Mars and Its Inhabitants."
28. Best, "Mars," 234; "Martian Gospel"; Freeman and Freeman, "I've a Sweetheart in Mars"; Norris, "A Good Time in Mars"; Rose and Snyder, "I've Just Had a Message;' Taylor, "A Signal from Mars"; Tennent, "The Beautiful Planet of Mars"; Mudie, "Mars"; Davis, "I Just Got a Message"; Johnson, "A Temperance Message from Mars"; Ganthony, *A Message from Mars*.
29. Ball, "Mars"; Lockyer, "Opposition of Mars"; and Pickering, "Signaling to Mars."
30. Brooks, "Signaling with Mars," 28. For cautions, see Ball, "Signaling to Mars," and Larkin, "Signaling to Mars."
31. Tesla, "Talking with Planets"; Tesla, "Signalling to Mars"; "Camille Flammarion's Latest Views," 137.
32. Markley, *Dying Planet*.
33. Bell, Butlin, and Heffernan, *Geography and Empire*; Driver, *Geography Militant*; Smith, *American Empire*; Schulten, *Geographical Imagination*; Livingstone, *Geographical Tradition*; Orde, *Eclipse of Great Britain*.
34. "Question of Life on Mars," 780–81.
35. Ibid., 782.
36. Clerke, "New Views about Mars," 384–85.
37. Wells, *War of the Worlds*.
38. Schroeder, "Message from Mars," 44.
39. See Heffernan, "Limits of Utopia."
40. Rhodes, *Last Will and Testament*, 190.
41. Markley, *Dying Planet*, 124. See also his discussion of dystopian Mars-related fiction works written by non-British Europeans, whose countries were dealing with many of the same imperial challenges.
42. Lowell's theory was supported in Wicks, "'Canals' of Mars—End of a Great Delusion"; Wicks, "Mars"; Wicks, "Photographs of Mars"; and Wicks, *To Mars via the Moon*.
43. Pickering, "Planet Mars," 469.
44. Lowell, *Mars*, 208–9.
45. Serviss, "Professor Lowell's Last Conclusions."
46. Burroughs, *A Princess of Mars*. His stories were first serialized in *All-Story Weekly* beginning in 1912.
47. Pfitzer, "Only Good Alien Is a Dead Alien," characterizes this novel as a derivative extension of the standard American frontier myth to outer space, in which the hero leaves civilization, earns respect among brutal savages, becomes a hero in his adopted society, and then returns home to Western civilization. Despite fundamental agreement that the narrative derives from a literary tradition based on "the metaphysics of Indian-hating" (cf. Drinnon, *Facing West*), I would caution that it is not quite as straightforward as Pfitzer asserts. The racial hierarchy of the Martian societies John Carter encounters defies an across-the-board characterization of Martian society and directly engages some of the more prominent concerns about cultural hierarchy that had been raised by Lowell's theory of civilization. To a certain extent, these new concerns could not be fully addressed within the old American frontier myth.
48. "Earth's Future," 170.
49. Lipset, *American Exceptionalism*; Goetzmann, *Exploration and Empire*; Livingstone, *Geographical Tradition*; John, "Benevolent Imperialism."
50. Brady, "Full of Empty"; Dunbar-Ortiz, *Roots of Resistance*; John, "Cultural Nationalism"; Bruckner, "Literacy for Empire"; Drinnon, *Facing West*; Frenkel, "Geography, Empire, and Environmental Determinism."

51. Smith, "Lost Geography," 12. See also Driver, "Commentary on Neil Smith."
52. Tuason, "Ideology of Empire." Schulten, *Geographical Imagination*, 13. See also Domosh, *American Commodities*.
53. Hegglund, "Empire's Second Take," 267; Said, *Culture and Imperialism*.
54. Said, *Orientalism*, 1, 4, 6; Schulten, *Geographical Imagination*, 13.
55. Burke, "Stereotypes of Others," 123.
56. Stern, *Eugenic Nation*; Fairchild, "Rise and Fall of the Medical Gaze"; Winlow, "Mapping Moral Geographies"; Kurashige, "Immigration, Race, and the Progressives."
57. MacDonald, "Anti-*Astropolitik*," 596, 597.
58. Kitchin and Neale, "Science Fiction or Future Fact"; Dittmer, "Colonialism and Place Creation"; Badescu, *Mars*. MacDonald, "Anti-*Astropolitik*," shows that Dolman's currently influential model of *Astropolitik* draws explicitly on classic texts in imperial geography, such as Mackinder, "Geographical Pivot of History," and Mahan, *Influence of Sea Power Upon History*.

BIBLIOGRAPHY

ARCHIVAL COLLECTIONS

British Library, London
Harvard College Observatory Archives, Cambridge, Massachusetts
Harvard University Houghton Library, Cambridge, Massachusetts
L'Archivo Storico dell'Osservatorio Astronomico di Brera, Milan, Italy
Lowell Observatory Archives, Flagstaff, Arizona
Royal Astronomical Society, London
Royal Geographical Society, London
Royal Greenwich Observatory Archives, Cambridge, England
U.S. Library of Congress, Washington, D.C.

PUBLISHED HISTORICAL MATERIALS

Agassiz, G. R. "The Markings on Mars." *The Nation* 86, no. 2220 (1908): 56.
———. "Mars as Seen in the Lowell Refractor." *Popular Science Monthly* 71 (1907): 275–82.
All'astronomo G. V. Schiaparelli: Omaggio 30 Giugno 1860–30 Giugno 1900. Milan: Osservatorio Astronomico di Brera, 1900.
Antoniadi, E. M. "Mars Section Fifth Interim Report for 1909, Dealing with the Fact Revealed by Observation That Prof. Schiaparelli's 'Canal' Network Is the Optical Product of the Irregular Minor Details Diversifying the Martian Surface." *Journal of the British Astronomical Association* 20, no. 3 (1909): 136–41.
———. "Mars Section Fourth Interim Report for the Apparition of 1909, Dealing with the Appearance of the Planet Mars between September 20 and October 23 in the Great Refractor of the Meudon Observatory." *Journal of the British Astronomical Association* 20, no. 2 (1909): 78–81.
———. "Mars Section Interim Report on the Australian Observations, 1907." *Journal of the British Astronomical Association* 18, no. 10 (1908): 398–401.
———. "Mars Section, Second Interim Report for 1898–99." *Journal of the British Astronomical Association* 9, no. 8 (1899): 367–71.
———. "Mars Section Sixth Interim Report for 1909, Dealing with Some Further Notes on the So-Called 'Canals.'" *Journal of the British Astronomical Association* 20, no. 4 (1910): 189–92.

————. "Mars Section Third Interim Report for 1909, Dealing with the Nature of the So-Called 'Canals' of Mars." *Journal of the British Astronomical Association* 20, no. 1 (1909): 25–28.

————. "Note on Some Photographic Images of Mars Taken in 1907 by Professor Lowell." *Monthly Notices of the Royal Astronomical Society* 69, no. 2 (1908): 110–14.

————. "On the Possibility of Explaining on a Geomorphic Basis the Phenomena Presented by the Planet Mars." *Journal of the British Astronomical Association* 20, no. 2 (1909): 89–94.

————. "Section for the Observation of Mars: Report of the Section, 1896." *Memoirs of the British Astronomical Association* 6 (1898): 55–102.

————. "Section for the Observation of Mars: Report of the Section, 1900–1901." *Memoirs of the British Astronomical Association* 11 (1903): 85–142.

————. "Section for the Observation of Mars: Report of the Section, 1909." *Memoirs of the British Astronomical Association* 20 (1916): 25–92.

"The Atmosphere of Mars." *The Observatory* 17, no. 219 (1894): 341–42.

Bailey, Solon I. "Astronomical Notes: The Planet Mars." *Science* n.s. 26, no. 678 (1907): 910–12.

————. "Expeditions and Foreign Stations." In *The History and Work of Harvard Observatory, 1839–1927: An Outline of the Origin, Development, and Researches of the Astronomical Observatory of Harvard College Together with Brief Biographies of Its Leading Members.* New York: McGraw-Hill, 1931.

————. "Harvard Observatory in Peru." *Scientific American* 76 (1897): 329–31.

————. *The History and Work of Harvard Observatory, 1839–1927: An Outline of the Origin, Development, and Researches of The Astronomical Observatory of Harvard College Together with Brief Biographies of Its Leading Members.* New York: McGraw-Hill, 1931.

————. "History of the Expedition." *Annals of the Astronomical Observatory of Harvard College* 34 (1895): 1–48.

Ball, Robert S. *In the High Heavens.* London: Isbister, 1893.

————. "Mars." *The Living Age* 195, no. 2519 (1892): 195–205.

————. "Signaling to Mars." *Scientific American Supplement,* June 8, 1901, 21267–68.

————. "Signalling to Mars." *The Living Age* 229, no. 2963 (1901): 277–84. .

Best, St. George. "Mars" (poem). *New England Magazine* 13, no. 2 (1892): 234.

Blackwelder, Eliot. "Mars as the Abode of Life." *Science* n.s. 29, no. 747 (1909): 659–61.

Brewster, E. T. "The Earth and the Heavens." *Atlantic Monthly* 100, no. 2 (1907): 260–65.

Brooks, William R. "Signaling with Mars." *Collier's* 44 (1909): 27–28.

Burroughs, Edgar Rice. *A Princess of Mars.* New York: Dover Publications, 1964.

Burton, C. E. "Canals on Mars." *Astronomical Register* 20, no. 234 (1882): 142.

"Camille Flammarion's Latest Views on Martian Signaling." *Scientific American Supplement,* August 31, 1907, 137.

Cammell, Bernard E. "Mars Section, 1894." *Journal of the British Astronomical Association* 4, no. 9 (1894): 395–97.

Campbell, W. W. "Concerning an Atmosphere on *Mars.*" *Publications of the Astronomical Society of the Pacific* 6, no. 38 (1894): 273–83.

————. "The Projections of the Planet Mars." *Scientific American* 89, no. 5 (1903): 82.

————. "The Spectrum of Mars." *Publications of the Astronomical Society of the Pacific* 6 (1894): 228–36.

————. "The Spectrum of Mars as Observed by the Crocker Expedition to Mt. Whitney." *Lick Observatory Bulletin* 169 (1909): 149–64.

"The Canals on Mars." *Astronomical Register* 24, no. 286 (1886): 268.

"Canals on Mars." *Scientific American Supplement*, January 27, 1906, 25143.

Chambers, George F. *A Handbook of Descriptive Astronomy*. 3rd ed. Oxford: Clarendon Press, 1877.

Clerke, Agnes M. "New Views about Mars." *Edinburgh Review* 184 (1896): 368–85.

———. *A Popular History of Astronomy During the Nineteenth Century*. London: A. and C. Black, 1908.

Crommelin, A.C.D. "Martian Photography." *The Observatory* 30, no. 387 (1907): 365.

Davis, Gussie L. "I Just Got a Message from Mars." Sheet music. New York: Howley, Haviland and Co., 1896.

Deakin, Alfred. *Irrigated India: An Australian View of India and Ceylon, Their Irrigation and Agriculture*. London: W. Thacker, 1893.

Douglass, Andrew Ellicott. "The Lowell Observatory and Its Work." *Popular Astronomy* 2, no. 9 (1895): 395–402.

———. "Mars." *Popular Astronomy* 7, no. 3 (1899): 113–17.

———. "Scales of Seeing." *Popular Astronomy* (1898).

"The Earth's Future." *Brown Alumni Monthly* 7, no. 8 (1907): 170.

Fennessy, E. B. "Nomenclature of Markings on Mars." *Astronomical Register* 17, no. 195 (1879): 70.

Flammarion, Camille. *La planète Mars, et ses conditions d'habitabilité. Synthese generale de toutes les observations. Climatologie, meteorologie, areographie, continents, mers, et rivages, eaux et neiges, saisons, variations observees*. Paris: Gauthier-Villars et Fils, 1892.

———. "Mars and Its Inhabitants." *North American Review* 162, no. 474 (1896): 546–77.

———. "Recent Observations of Mars." *Scientific American* 74 (1896): 133–34.

Fleming, M. "Harvard College Observatory Astronomical Expedition to Peru." *Publications of the Astronomical Society of the Pacific* 4 (1892): 58–62.

Freeman, Frederick W., and Clara Gregg Freeman. "I've a Sweetheart in Mars." Sheet music. Chicago: Victor Kremer, 1908.

"French Clergyman Combats Theory of Prof. Lowell as to Presence of Some Sort of Intelligent Life on Planet Mars." *Sunday Herald*, August 4, 1907, magazine section.

Ganthony, Richard. *A Message from Mars* (dramatic play). London: Staged at the Avenue Theatre, 1899.

Gernsback, Hugo. "How the Martian Canals Are Built." *Electrical Experimenter*, November 1916, 486–87, 513, 539.

Gill, David. "The Nomenclature of Markings on Mars." *Astronomical Register* 17, no. 196 (1879): 95.

Green, Nathaniel E. "The Approaching Opposition of Mars." *Monthly Notices of the Royal Astronomical Society* 39, no. 8 (1879): 433.

———. "The Northern Hemisphere of Mars." *Monthly Notices of the Royal Astronomical Society* 46, no. 8 (1886): 445–47.

———. "Notes on the Coming Opposition of Mars." *Monthly Notices of the Royal Astronomical Society* 37, no. 7 (1877): 424.

———. "Observations of Mars, at Madeira, in August and September 1877." *Memoirs of the Royal Astronomical Society* 44 (1879): 123–40.

Gregory, R. A. "Mars as a World." *The Living Age* 225, no. 2909 (1900): 21–28.

Hale, Edward E. "Latest News from Mars." *Publications of the Astronomical Society of the Pacific* 7 (1895): 116–18.

———. "Latest News from Mars." *Scientific American* 72, no. 9 (1895): 137.

Hale, George E. "The Aim of the Yerkes Observatory." *Astrophysical Journal* 6 (1897): 310–21.

————. "The Development of a New Observatory." *Publications of the Astronomical Society of the Pacific* 17 (1905): 41–52.

————. "The Yerkes Observatory of the University of Chicago: 1. Selection of the Site." *Astrophysical Journal* 5 (1897): 164–80.

"Harvard Observatory in Peru—the Highest Meteorological Station in the World." *Scientific American* 70 (1894): 67.

Herschel, John F. W. *Outlines of Astronomy*. 10th ed. London: Longmans, Green, 1875.

Heward, E. Vincent. "Mars: Is It a Habitable World?" *The Living Age* 254, no. 3298 (1907): 741–51.

Hoffman, Otto. "The Canals of Mars." *Scientific American Supplement*, December 3, 1910, 362.

Holden, Edward S. *A Brief Account of the Lick Observatory of the University of California*. 2nd ed. Sacramento, Calif.: State Publishing Office, 1895.

————. "The Latest News of Mars." *North American Review* 160, no. 462 (1895): 636–38.

————. "The Lick Observatory." *New York Daily Tribune*, September 2, 1888, 3.

————. "The Lick Observatory." *Sidereal Messenger* 3 (1884): 301–3.

————. "The Lick Observatory." *Sidereal Messenger* 7 (1888): 47–65.

————. "The Lowell Observatory, in Arizona." *Publications of the Astronomical Society of the Pacific* 6 (1894): 160–69.

————. *Mountain Observatories in America and Europe*. Washington, D.C.: Smithsonian Institution, 1896.

————. "Note on the Mount Hamilton Observations of Mars, June–August 1892." *Astronomy and Astro-physics* 11, no. 108 (1892): 663–68.

————. "What We Really Know about Mars." *The Forum* 14 (1892): 359–68.

Hollings, Mary Albright. *The Life of Sir Colin C. Scott-Moncrieff*. London: J. Murray, 1917.

Holmes, Edwin. "The Canals of Mars." *Journal of the British Astronomical Association* 10, no. 7 (1900): 300–304.

————. "Communication with Mars." *Journal of the British Astronomical Association* 11, no. 5 (1901): 202–6.

————. "Notes re Mars." *Journal of the British Astronomical Association* 1, no. 5 (1891): 256–59.

Holt, J. R. "The Solar Image Reflected in the Seas of Mars." *Astronomy and Astro-physics* 13, no. 123 (1894): 257–58.

Housden, C. E. *Riddle of Mars: The Planet*. London: Longmans, Green, 1914.

Huntington, Ellsworth. *Civilization and Climate*. New Haven: Yale University Press, 1915.

"An Image of the Sun on the Martian Seas." *Journal of the British Astronomical Association* 4, no. 6 (1894): 260–61.

"In Memoriam: Nathaniel E. Green, F.R.A.S." *Journal of the British Astronomical Association* 10, no. 2 (1899): 75–77.

Johnson, Bob. "A Temperance Message from Mars (?)." Sheet music. In *The Standard Book of Song for Temperance Meetings and Home Use*. London: National Temperance Publication Depot, 1901.

Kaempffert, Waldemar. "What We Know about Mars." *McClure's Magazine* 28, no. 5 (1907): 481–86.

Keeler, James E. "Picturing the Planets: Portraits of Jupiter, Mars, and Saturn, and How They Were Made at the Lick Observatory." *Century Magazine* 50, no. 3 (1895): 455–62.

Larkin, Edgar Lucien. "Signaling to Mars: Its Impossibility by Means of Light." *Scientific American Supplement*, June 19, 1909, 387.

Lepper, G. H. "An Examination of the Modern Views as to the Real Nature of the Markings of Mars." *Journal of the British Astronomical Association* 15, no. 3 (1905): 133–37.

"The Lick Observatory." *Sidereal Messenger* 4 (1885): 46–49.

"The Lick Observatory of the University of California." *Scientific American* 58 (1888): 159–63.

"Life at the Lick Observatory." *Scientific American* 64 (1891): 73.

"Life in Mars." *Chambers's Journal of Popular Literature, Science, and Art* 3 (5th ser.), no. 128 (1886): 369–71.

Lockyer, J. Norman. *Elementary Lessons in Astronomy.* London: Macmillan, 1894.

———. "The Opposition of Mars." *Nature: A Weekly Illustrated Journal of Science* 46, no. 1193 (1892): 443–48.

Lowell, Percival. "Cartouches of the Canals of Mars." *Bulletins of the Lowell Observatory* 1, no. 12 (1904): 59–86.

———. *Chosön: The Land of the Morning Calm: A Sketch of Korea.* Boston: Ticknor, 1886.

———. "Explanation of the Supposed Signals from Mars of December 7, and 8, 1900." *Popular Astronomy* 10, no. 4 (1902): 185–94.

———. *Mars.* Boston: Houghton, Mifflin, 1895.

———. "Mars." *Popular Astronomy* 2, no. 1 (1894): 1–8.

———. *Mars and Its Canals.* New York: Macmillan, 1906.

———. *Mars as the Abode of Life.* New York: Macmillan, 1908.

———. "Mars: The Flagstaff Photographs." *New England Magazine* n.s.12, no. 6 (1895): 643–54.

———. "Mars: Atmosphere." *Atlantic Monthly* 75, no. 451 (1895): 594–603.

———. "Mars: Canals." *Atlantic Monthly* 76, no. 453 (1895): 106–19.

———. "Mars: Oases." *Atlantic Monthly* 76, no. 454 (1895): 223–35.

———. "Mars: Oases." *Popular Astronomy* 2, no. 8 (1895): 343-48.

———. "Mars: Seasonal Changes on the Planet's Surface." *Astronomy and Astro-physics* 13, no. 130 (1894): 814–21.

———. "Mars: Spring Phenomena." *Popular Astronomy* 2, no. 3 (1894): 97–100.

———. "Mars: The Canals. I." *Popular Astronomy* 2, no. 6 (1895): 255–61.

———. "Mars: The Polar Snows." *Popular Astronomy* 2, no. 2 (1894): 52–56.

———. "Mars: The Water Problem." *Atlantic Monthly* 75, no. 452 (1895): 749–58.

———. "The New Canals of Mars: Recent Discoveries at Flagstaff." *Scientific American Supplement*, March 19, 1910, 190–91.

———. "New Photographs of Mars: Taken by the Astronomical Expedition to the Andes and Now First Published." *Century Magazine* 75 (1907): 303–11.

———. *Noto: An Unexplored Corner of Japan.* Boston: Houghton, Mifflin, 1891.

———. "On the Climatic Causes of the Removal of the Lowell Observatory to and from Mexico." *The Observatory* 20, no. 259 (1897): 401–4.

———. "The Planet Mars." *McClure's Magazine* 30, no. 2 (1907): 223–36.

———. "The Plateau of the San Francisco Peaks in Its Effect on Tree-Life. Part I." *Bulletin of the American Geographical Society* 41 (1909): 257–70.

———. "The Plateau of the San Francisco Peaks in Its Effect on Tree-Life. Part II (Conclusion)." *Bulletin of the American Geographical Society* 41 (1909): 365–82.

———. "Schiaparelli." *Popular Astronomy* 18, no. 8 (1910): 456–67.

———. *The Soul of the Far East.* Boston: Houghton Mifflin, 1888.

———. "The Spectra of the Major Planets." *Nature* 79, no. 2037 (1908): 42.

———. "The Thoth and the Amenthes." *Bulletins of the Lowell Observatory* 1, no. 8 (1904): 39–43.

Mackinder, Halford. "The Geographical Pivot of History." *Geographical Journal* 23 (1904): 421–37.

Maggini, Mentore. *Il Pianeta Marte*. Milan: Librario Ulrico Hoepli, 1930.

Mahan, A. *The Influence of Sea Power Upon History, 1660–1783*. Boston: Little, Brown, 1890.

"The Man Who Made the Martians Live." *Lawrence* (Mass.) *Evening Tribune*, November 15, 1916.

Manson, Marsden. "The Climate of Mars." *Popular Astronomy* 2, no. 8 (1895): 371–74.

"Mapping the Southern Sky from a Mountain Peak 14,000 Feet High." *Scientific American* 64 (1891): 36.

"Mars and His Moons." *Scribner's Monthly: An Illustrated Magazine for the People* 15, no. 2 (1877): 263–66.

"The Mars Man." *Paterson* (N.J.) *Guardian*, December 1, 1916.

Marth, A. "Nomenclature of Markings Visible Upon the Planet Mars." *Astronomical Register* 17, no. 193 (1879): 24–25.

"A Martian Gospel." *The Nation* 84, no. 2191 (1907): 583–84.

Maunder, E. Walter. "Mars Section." *Journal of the British Astronomical Association* 2, no. 9 (1892): 480–81.

———. "Section for the Observation of Mars: Report of the Section, 1892." *Memoirs of the British Astronomical Association* 2 (1895): 157–98.

Mee, Arthur. *Observational Astronomy, a Book for Beginners*. Cardiff: Daniel Owen, 1893.

"Meeting of the Royal Astronomical Society, April 12, 1878." *Astronomical Register* 16, no. 185 (1878): 115–23.

"Meeting of the Royal Astronomical Society, April 14, 1882." *Astronomical Register* 20, no. 233 (1882): 101–11.

"Meeting of the Royal Astronomical Society, December 13, 1878." *Astronomical Register* 17, no. 193 (1879): 1–20.

"Meeting of the Royal Astronomical Society, November 8, 1877." *Astronomical Register* 15, no. 180 (1877): 309–19.

Montgomery, George Edgar. "Mars" (poem). *Harper's New Monthly Magazine* 93, no. 556 (1896): 611.

Morse, Edward S. *Mars and Its Mystery*. Boston: Little, Brown, 1906.

———. "My 34 Nights on Mars: How Prof. Edward S. Morse Has Been Studying the Great Planet through the Lowell Observatory Telescope and His Own Interesting Account of What He Discovered There." *The World Magazine*, October 7, 1906, 9.

Mudie, Walter. "Mars." Sheet music. In *Child Thoughts*. London: Forsyth Brothers, 1920.

Newcomb, Simon. "Astronomy." In *The Encyclopaedia Britannica, 11th Edition*, ed. Hugh Chisholm, 800–819. Cambridge: University Press, 1910.

———. "Fallacies about Mars." *Harper's Weekly* 52 (1908): 11–12.

———. *His Wisdom the Defender*. New York: Harper and Brothers, 1900.

———. "Mars." In *Johnson's Universal Cyclopaedia*, ed. Charles Kendall Adams, 571. London: D. Appleton, 1893.

———. "The Optical and Psychological Principles Involved in the Interpretation of the So-Called Canals of Mars." *Astrophysical Journal* 26, no. 1 (1907): 1–17.

Newcomb, Simon, and Edward S. Holden. *Astronomy for Schools and Colleges*. New York: Henry Holt, 1879.

Noble, William. "Names of Markings on Mars." *Astronomical Register* 17, no. 196 (1879): 95–96.

Norris, Harry B. "A Good Time in Mars." Sheet music. London: Francis, Day, and Hunter, 1905.

"Notes." *The Observatory* 31, no. 396 (1908): 221.

"Notes: The Opposition of Mars." *Journal of the British Astronomical Association* 2, no. 9 (1892): 477.

Orr, M. A. [Mary Acworth]. "The Canals of Mars." *Knowledge: An Illustrated Magazine of Science* 24 (1901): 38–39.

Osservatorio Astronomico di Brera. *Corrispondenza Su Marte Di Giovanni Virginio Schiaparelli*. Vol. 1. Pisa: Domus Galilaeana, 1963.

Osservatorio Astronomico di Brera. *Corrispondenza Su Marte Di Giovanni Virginio Schiaparelli*. Vol. 2. Pisa: Domus Galilaeana, 1976.

"The Past Opposition of Mars. In the Report of the Council to the Seventy-Seventh Annual General Meeting of the Society." *Monthly Notices of the Royal Astronomical Society* 57, no. 4 (1897): 284–86.

Payne, W. W. "The 'Canals' of Mars." *Popular Astronomy* 12, no. 6 (1904): 365–75.

Perrine, C. D. "List of Earthquakes in California for the Years 1891–1892." *Publications of the Astronomical Society of the Pacific* 5 (1893): 127–30.

Perrotin, J. "Observation des canaux de Mars faite a l'observatoire de Nice." *The Observatory* 9, no. 116 (1886): 364–65.

Pickering, Edward C. "A Climb in the Cordillera of the Andes." *Appalachia* 7 (1894): 205–12.

———. "Colors Exhibited by the Planet Mars." *Astronomy and Astro-physics* 11, no. 106 (1892): 449–53.

———. "Colors Exhibited by the Planet Mars." *Astronomy and Astro-physics* 11, no. 107 (1892): 545–48.

———. "An Explanation of the Martian and Lunar Canals." *Popular Astronomy* 12, no. 7 (1904): 439–42.

———. "Mars." *Astronomy and Astro-physics* 11, no. 108 (1892): 668–75.

———. "Mountain Observatories." *Appalachia* 3 (1883): 99–106.

———. "The Planet Mars." *Technical World Magazine* (1906): 459–71.

———. "The Seas of Mars." *Astronomy and Astro-physics* 13, no. 127 (1894): 553–56.

———. "Signaling to Mars." *Scientific American* 100 (1909): 43.

"The Planet Mars." *The Observatory* 15, no. 194 (1892): 413–14.

"The Planet Mars." *Popular Astronomy* 15, no. 7 (1907): 449–50.

Powell, J. W. *Report on the Lands of the Arid Region of the United States*. 2nd ed. Washington, D.C.: Government Printing Office, 1879.

Proctor, Richard A. "Names of Markings on Mars." *Astronomical Register* 17, no. 194 (1879): 45–46.

———. "Note on Mars." *Monthly Notices of the Royal Astronomical Society* 48, no. 6 (1888): 307–08.

———. *The Orbs Around Us: A Series of Familiar Essays on the Moon and Planets, Meteors and Comets, the Sun and Coloured Pairs of Suns*. London: Longmans, Green, 1872.

———. *Other Worlds Than Ours: The Plurality of Worlds Studied Under the Light of Recent Scientific Researches*. New York: D. Appleton, 1871.

———. *Other Worlds Than Ours: Part One—Science*. In *A Library of Universal Literature*. New York: P. F. Collier and Son, 1900.

"Publications: *Mars*." *The Observatory* 19, no. 242 (1896): 280–81.

"The Question of Life on Mars." *The Living Age* 258, no. 3351 (1908): 771–86.

Ratzel, Friedrich. *The History of Mankind [Völkerkunde]*. 3 vols. London: Macmillan, 1896–98.

"Report of the Meeting of the Association, Held on June 20, 1906, at Sion College, Victoria Embankment." *Journal of the British Astronomical Association* 16, no. 9 (1906): 333.

"Report of the Meeting of the Association, Held on June 24, 1903, at Sion College, Victoria Embankment." *Journal of the British Astronomical Association* 13, no. 9 (1903): 331–40.

"Report of the Meeting of the Association, Held on March 30, 1910, at Sion College, Victoria Embankment, E.C." *Journal of the British Astronomical Association* 20, no. 6 (1910): 285–94.

"Report of the Meeting of the Association, Held on Wednesday, December 29, 1909, at Sion College, Victoria Embankment, E.C." *Journal of the British Astronomical Association* 20, no. 3 (1909): 119–28.

"Report on the Meeting of the Association Held December 31, 1890." *Journal of the British Astronomical Association* 1, no. 3 (1890): 110–14.

Reynold's Universal Atlas of Astronomy, Geology, Physical Geography, the Vegetable Kingdom, and Natural Philosophy, Comprising Four Hundred Coloured Maps and Diagrams, with Popular Descriptions. London: James Reynolds, 1876.

Rhodes, Cecil. *The Last Will and Testament of Cecil John Rhodes: With Elucidatory Notes to Which Are Added Some Chapters Describing the Political and Religious Ideas of the Testator*. London: "Review of Reviews" Office, 1902.

"The Riddle of Mars: How the Planet May Be Saving Itself from Death by Irrigation." *Scientific American Supplement*, August 15, 1914, 106–7.

Rose and Snyder. "I've Just Had a Message from Mars." Sheet music. New York: Howley, Haviland and Dresser, 1903.

Schiaparelli, Giovanni Virginio. *Astronomical and Physical Observations of the Axis of Rotation and the Topography of the Planet Mars: First Memoir, 1877–1878*. Translated by William Sheehan, A.L.P.O. Monographs: Association of Lunar and Planetary Observers, 1996.

———. "Gli abitanti di altri mondi." In *Le opere di G.V. Schiaparelli*, vol. 10, ed. Reale Specola di Brera, 122–25. Milano: Ulrico Hoepli, 1940.

———. "Osservazioni astronomiche e fisiche sull'asse di rotazione e sulla topografia del pianeta Marte fatte nella reale specola di Brera in Milano coll'equatoreale di Merz : Memoria seconda del socio G.V. Schiaparelli." *Atti della Reale Accademia dei Lincei: Memorie della classe di scienze fisiche, matematiche e naturali* 3, no. 10 (1880): 3–109.

———. "Osservazioni astronomiche e fisiche sull'asse di rotazione e sulla topografia del pianeta Marte fatte nella reale specola di Brera in Milano coll'equatoreale di Merz : Memoria terza del socio G.V. Schiaparelli (opposizione 1881–1882)." *Atti della Reale Accademia dei Lincei: Memorie della classe di scienze fisiche, matematiche e naturali* 4, no. 3 (1886): 281–373.

———. "Osservazioni astronomiche e fisiche sull'asse di rotazione e sulla topografia del pianeta Marte fatte nella reale specola di Brera in Milano coll'equatoreale di Merz durante l'opposizione del 1877: Memoria del socio G.V. Schiaparelli." *Atti della Reale Accademia dei Lincei: Memorie della classe di scienze fisiche, matematiche e naturali* 3, no. 2 (1877–78): 3–136.

———. "Osservazioni Astronomiche e fisiche sull'asse di rotazione e sulla topografia del pianeta Marte fatte nella reale specola di Brera in Milano coll'equatoreale di Merz: Memoria quarta del socio G.V. Schiaparelli." *Atti della Reale Accademia dei Lincei: Memorie della classe di scienze fisiche, matematiche e naturali* 5, no. 2 (1895): 183–240.

———. "The Planet Mars." *Astronomy and Astro-physics* 13, no. 128 (1894): 635–40.

———. "The Planet Mars." *Astronomy and Astro-physics* 13, no. 129 (1894): 714–23.

———. "Topografia e clima di Milano." In *Le opere di G.V. Schiaparelli*, vol. 11, ed. Reale Specola di Brera, 355–96. Milano: Ulrico Hoepli, 1943.

"Schiaparelli's Observations of Mars." *The Observatory* 5, no. 61 (1882): 138–43.

"Science." *The Nation* 84, no. 2179 (1907): 317–18.

"Scientists Will Try to Solve the Martian Riddle: Lowell Expedition Will Photograph Planet from Lofty Andes Peaks." *Philadelphia Inquirer*, May 13, 1907, 1.

Semple, Ellen Churchill. *American History and Its Geographic Conditions*. Boston: Houghton Mifflin, 1903.

Serviss, Garrett P. "Professor Lowell's Last Conclusions about Life on Mars." *New York City American*, December 10, 1916.

"The Signals from Mars." *Popular Astronomy* 3, no. 1 (1895): 47.

Slipher, E. C. "Photographing Mars." *Century Magazine* 75 (1907): 312.

Smythe, William E. *The Conquest of Arid America*. New York: Macmillan, 1905.

"'Something Like Human Intelligence on Mars': Observations Made by Professor Todd in the Andes Give Rise to Interesting Theories of Life on Neighboring Planet." *New York Times*, October 27, 1907, SM3.

Taylor, Raymond. "A Signal from Mars: March and Two Step." Sheet music. New York: E. T. Paull Music Co., 1901.

Tennent, H. M. "The Beautiful Planet of Mars." Sheet music. London: Chappell, 1911.

Terby, F. "Nomenclature of Martial Markings." *Astronomical Register* 17, no. 194 (1879): 46–47.

Tesla, Nikola. "Signalling to Mars—A Problem of Electrical Engineering." *Harvard Illustrated Magazine*, March 1907, 119–21.

———. "Talking with Planets." *Collier's Weekly* (1901): 4–5.

"To Photograph Mars: Lowell Party to Take Views on Andes Range with a Huge Camera." *New York Times*, May 12, 1907, 1.

Todd, David. "The Lowell Expedition to the Andes." *Popular Astronomy* 15, no. 9 (1907): 551–53.

———. "Professor Todd's Own Story of the Mars Expedition: First Article Published from the Pen of the Leader of the Party of Observation." *Cosmopolitan Magazine* 44, no. 4 (1908): 343–51.

Todd, Mabel Loomis. "A Great Modern Observatory: Harvard's Astronomical Work." *Century Magazine* 54, no. 2 (1897): 290–300.

———. "Our Ruddy Neighbor Planet." *The Independent* 64, no. 3097 (1908): 791–95.

———. "Photographing the 'Canals' on Mars." *The Nation* 85, no. 2203 (1907): 264–66.

Turner, Frederick Jackson. "The Significance of the Frontier in American History." In *The Frontier in American History*, 1–38. 1893; Huntington, N.Y.: Robert E. Krieger Publishing, 1976.

Wallace, Alfred Russel. *Is Mars Habitable? A Critical Examination of Professor Percival Lowell's Book "Mars and Its Canals," with an Alternative Explanation*. London: Macmillan, 1907.

———. *Man's Place in the Universe: A Study of the Results of Scientific Research in Relation to the Unity or Plurality of Worlds*. New York: Phillips, 1903.

———. *The Wonderful Century: Its Successes and Its Failures*. New York: Dodd, Mead, 1898.

Ward, Lester Frank. "Mars and Its Lesson." *Brown Alumni Monthly* 7, no. 8 (1907): 159–65.

Warner, Charles Dudley. "Editor's Study." *Harper's New Monthly Magazine* 93, no. 556 (1896): 635–40.

Wells, H. G. "The Things That Live on Mars: A Description, Based Upon Scientific Reasoning, of the Flora and Fauna or Our Neighboring Planet, in Conformity with the Very Latest Astronomical Revelations." *Cosmopolitan Magazine* 44, no. 4 (1908): 335–42.

———. *The War of the Worlds.* London: William Heinemann, 1898.

"Why the Dwellers on Mars Do Not Make War." *Current Literature* 42, no. 2 (1907): 211–14.

Wicks, Mark. "The 'Canals' of Mars—The End of a Great Delusion" (letter). *English Mechanic and World of Science* 82, no. 2119 (1905): 298.

———. "The Canals of Mars" (letter). *English Mechanic and World of Science* 82, no. 2124 (1905): 403.

———. "Mars: To Mr. T. K. Mellor and Others" (letter). *English Mechanic and World of Science* 82, no. 2126 (1905): 450–51.

———. "Photographs of Mars" (letter). *English Mechanic and World of Science* 82, no. 2129 (1906): 516–17.

———. *To Mars via the Moon: An Astronomical Story.* London: Seeley, 1911.

"Will the New Year Solve the Riddle of Mars?" *New York Herald,* December 30, 1906.

Willcocks, William. *Sixty Years in the East.* Edinburgh: W. Blackwood, 1935.

Wilson, H. C. "Mars and His Canals." *Sidereal Messenger* 8, no. 1 (1889): 13–25.

SECONDARY SOURCES

Aitken, Robert G. "William Wallace Campbell, 1862–1938." *Publications of the Astronomical Society of the Pacific* 50 (1938): 204–9.

Arnold, David, and Ramachandra Guha. *Nature, Culture, Imperialism: Essays on the Environmental History of South Asia.* Delhi: Oxford University Press, 1995.

Badescu, Viorel, ed. *Mars: Prospective Energy and Material Resources.* New York: Springer, 2010.

Barnes, Trevor. "History and Philosophy of Geography: Life and Death, 2005–2007." *Progress in Human Geography* 32, no. 5 (2008): 650–58.

Bassett, K. "'Whatever Happened to the Philosophy of Science?': Some Comments on Barnes." *Environment and Planning A* 25 (1994): 337–42.

Bassett, Thomas J., and Philip W. Porter. "'From the Best Authorities': The Mountains of Kong in the Cartography of West Africa." *Journal of African History* 32, no. 3 (1991): 367–413.

Bell, Morag, Robin Butlin, and Michael Heffernan. *Geography and Imperialism 1820–1940.* Edited by John M. MacKenzie, Studies in Imperialism. Manchester: Manchester University Press, 1995.

Berton, Pierre. *The Arctic Grail: The Quest for the North West Passage and the North Pole, 1818–1909.* New York: Viking, 1988.

Biagioli, Mario. *Galileo, Courtier: The Practice of Science in the Culture of Absolutism.* Chicago: University of Chicago Press, 1993.

Bishop, Peter. *The Myth of Shangri-La: Tibet, Travel Writing and the Western Creation of Sacred Landscape.* London: Athlone Press, 1989.

Blaut, J. M. *The Colonizer's Model of the World: Geographical Diffusionism and Eurocentric History.* New York: Guilford Press, 1993.

Bloom, Lisa. *Gender on Ice: American Ideologies of Polar Expeditions.* Minneapolis: University of Minnesota Press, 1993.

Blunck, Jurgen. *Mars and Its Satellites: A Detailed Commentary on the Nomenclature.* Hicksville, N.Y.: Exposition Press, 1977.

Blunt, Alison. "Imperial Geographies of Home: British Domesticity in India, 1886–1925." *Transactions of the Institute of British Geographers* 24 (1999): 421–40.

———. *Travel, Gender, and Imperialism: Mary Kingsley and West Africa.* New York: Guilford, 1994.

Boelhower, William. "Inventing America: A Model of Cartographic Semiosis." *Word and Image* 4, no. 2 (1988): 475–97.

Bourguet, Marie-Noelle, Christian Licoppe, and H. Otto Sibum. *Instruments, Travel and Science: Itineraries of Precision from the Seventeenth to the Twentieth Century.* London: Routledge, 2002.

Brady, Mary Pat. "'Full of Empty': Creating the Southwest as 'Terra Incognita.'" In *Nineteenth-Century Geographies: The Transformation of Space from the Victorian Age to the American Century,* ed. Helena Michie and Ronald R. Thomas, 251–64. New Brunswick, N.J.: Rutgers University Press, 2003.

Brantlinger, Patrick. "Victorians and Africans: The Genealogy of the Myth of the Dark Continent." *Critical Inquiry* 12 (1985): 166–202.

Broks, Peter. *Understanding Popular Science.* Berkshire, England: Open University Press, 2006.

Brown, Matthew. "Richard Vowell's Not-So-Imperial Eyes: Travel Writing and Adventure in Nineteenth-Century Hispanic America." *Journal of Latin American Studies* 38 (2006): 95–122.

Bruckner, Martin. "Literacy for Empire: The ABCs of Geography and the Rule of Territoriality in Early Nineteenth-Century America." In *Nineteenth-Century Geographies: The Transformation of Space from the Victorian Age to the American Century,* ed. Helena Michie and Ronald R. Thomas, 172–90. New Brunswick, N.J.: Rutgers University Press, 2003.

Brush, Stephen G. "A Geologist among Astronomers: The Chamberlin-Moulton Theory." In *Fruitful Encounters: The Origin of the Solar System and of the Moon from Chamberlin to Apollo,* 22–67. Vol. 3 of A History of Modern Planetary Physics. Cambridge: Cambridge University Press, 1996.

Burke, Peter. "Stereotypes of Others." In *Eyewitnessing: The Uses of Images as Historical Evidence,* 123–39. London: Reaktion Books, 2001.

Callon, Michel. "Some Elements of a Sociology of Translation: Domestication of the Scallops and the Fishermen of St. Brieuc Bay." In *The Science Studies Reader,* ed. Mario Biagioli, 67–83. New York: Routledge, 1999.

Cameron, Ardis. *Looking for America: The Visual Production of Nation and People.* Malden, Mass.: Blackwell, 2005.

Cantor, Geoffrey, and Sally Shuttleworth, eds. *Science Serialized: Representations of the Sciences in Nineteenth-Century Periodicals.* Cambridge, Mass.: MIT Press, 2004.

Chorley, R. J. "A Re-Evaluation of the Geomorphic System of W. M. Davis." In *Frontiers in Geographical Teaching,* ed. R. J. Chorley and Peter Haggett, 21–28. London: Methuen, 1965.

Coleman, William. "Science and Symbol in the Turner Frontier Hypothesis." *American Historical Review* 72, no. 1 (1966): 22–49.

Collins, H. M. *Changing Order: Replication and Induction in Scientific Practice.* London: Sage Publications, 1985.

———. "The TEA Set: Tacit Knowledge and Scientific Networks." In *The Science Studies Reader,* ed. Mario Biagioli, 95–109. New York: Routledge, 1999.

Cosgrove, Denis. *Apollo's Eye: A Cartographic Genealogy of the Earth in the Western Imagination.* Baltimore: Johns Hopkins University Press, 2001.

———. *Mappings.* London: Reaktion Books, 1999.

———. *Social Formation and Symbolic Landscape.* Madison: University of Wisconsin Press, 1984.

Cosgrove, Denis, and Stephen Daniels, eds. *The Iconography of Landscape: Essays on the Symbolic Representation, Design, and Use of Past Environments.* Cambridge: Cambridge University Press, 1988.

Cosgrove, Denis, and Veronica della Dora, eds. *High Places: Cultural Geographies of Mountains, Ice and Science.* London: IB Tauris, 2008.

Crowe, Michael J. *The Extraterrestrial Life Debate 1750–1900: The Idea of a Plurality of Worlds from Kant to Lowell.* Cambridge: Cambridge University Press, 1986.

Dartnell, Lewis. "A Living Mars?" *Geology Today* 24, no. 2 (2008): 62–67.

David, Robert G. *The Arctic in the British Imagination, 1818–1914.* Edited by John M. MacKenzie, Studies in Imperialism. Manchester: Manchester University Press, 2000.

Davis, Diana K. "Environmentalism as Social Control? An Exploration of the Transformation of Pastoral Nomadic Societies in French Colonial North Africa." *Arab World Geographer* 3, no. 3 (2000): 182–98.

———. "Potential Forests: Degradation Narratives, Science, and Environmental Policy in Protectorate Morocco, 1912–1956." *Environmental History* 10 (2005): 212–38.

———. *Resurrecting the Granary of Rome: Environmental History and French Colonial Expansion in North Africa.* Edited by James L.A. Webb Jr., Series in Ecology and History. Athens: Ohio University Press, 2006.

Davis, Mike. *Late Victorian Holocausts: El Nino Famines and the Making of the Third World.* New York: Verso Books, 2001.

DeVorkin, David H. "W. W. Campbell's Spectroscopic Study of the Martian Atmosphere." *Quarterly Journal of the Royal Astronomical Society* 18 (1977): 37–53.

Diamond, Jared. *Collapse: How Societies Choose to Fail or Succeed.* New York: Viking Books, 2005.

Dick, Steven J. *The Biological Universe: The Twentieth-Century Extraterrestrial Life Debate and the Limits of Science.* Cambridge: Cambridge University Press, 1996.

———. *Plurality of Worlds: The Origins of the Extraterrestrial Life Debate from Democritus to Kant.* Cambridge: Cambridge University Press, 1982.

Dittmer, Jason N. "Colonialism and Place Creation in *Mars Pathfinder* Media Coverage." *Geographical Review* 97, no. 1 (2007): 112–30.

Dobbins, Thomas A., and William Sheehan. "The Canals of Mars Revisited." *Sky and Telescope* 107, no. 3 (2004): 114–17.

Dolan, David Sutton. "Percival Lowell: The Sage as Astronomer." PhD diss., University of Wollongong, 1992.

Dolman, E. *Astropolitik: Classical Geopolitics in the Space Age.* London: Frank Cass, 2002.

Domosh, Mona. *American Commodities in an Age of Empire.* New York: Routledge, 2006.

Drinnon, Richard. *Facing West: The Metaphysics of Indian-Hating and Empire Building.* New York: New American Library, 1980.

Dritsas, Lawrence. "From Lake Nyassa to Philadelphia: A Geography of the Zambesi Expedition, 1858–64." *British Journal for the History of Science* 38 (2005): 35–52.

Driver, Felix. "Commentary on Neil Smith, *American Empire: Roosevelt's Geographer and the Prelude to Globalisation.*" *Political Geography* 24 (2005): 251–55.

———. "Distance and Disturbance: Travel, Exploration and Knowledge in the Nineteenth Century." *Transactions of the Royal Historical Society* 14 (2004): 73–92.

———. "Editorial: Field-Work in Geography." *Transactions of the Institute of British Geographers* 25, no. 3 (2000): 267–68.

———. "Exploration by Warfare: Henry Morton Stanley and His Critics." In *Geography Militant: Cultures of Exploration and Empire*, 117–45. Oxford: Blackwell, 2001.

———. "Geography, Empire and Visualisation: Making Representations." *Royal Holloway, University of London, Department of Geography Research Papers* General Series, no. 1 (1994): 1–17.

———. *Geography Militant: Cultures of Exploration and Empire.* Oxford: Blackwell, 2001.

———. "Geography's Empire: Histories of Geographical Knowledge." *Environment and Planning D: Society and Space* 10 (1992): 23–40.

———. "Henry Morton Stanley and His Critics: Geography, Exploration, and Empire." *Past and Present* 133 (1991): 134–66.

———. "Making Space." *Ecumene* 1 (1994): 386–90.

Driver, Felix, and Luciana Martins, eds. *Tropical Visions in an Age of Empire.* Chicago: University of Chicago Press, 2005.

Dunbar-Ortiz, Roxanne. *Roots of Resistance: A History of Land Tenure in New Mexico.* Norman: University of Oklahoma Press, 2007.

Duncan, James S., and Derek Gregory. *Writes of Passage: Reading Travel Writing.* London: Routledge, 1999.

Edgerton, Samuel Y. "From Mental Matrix to *Mappamundi* to Christian Empire: The Heritage of Ptolemaic Cartography in the Renaissance." In *Art and Cartography*, ed. David Woodward, 10–50. Chicago: University of Chicago Press, 1987.

Edney, Matthew H. *Mapping an Empire: The Geographical Construction of British India, 1765–1843.* Chicago: University of Chicago Press, 1997.

Endfield, Georgina H., and David J. Nash. "Missionaries and Morals: Climatic Discourse in Nineteenth-Century Central Southern Africa." *Annals of the Association of American Geographers* 92, no. 4 (2002): 727–42.

Fairchild, Amy L. "The Rise and Fall of the Medical Gaze: The Political Economy of Immigrant Medical Inspection in Modern America." *Science in Context* 19 (2006): 337–56.

Fan, Fa-Ti. "Victorian Naturalists in China: Science and Informal Empire." *British Journal for the History of Science* 36, no. 1 (2003): 1–26.

Fay, Charles E., Allen H. Bent, Howard Palmer, James M. Thorington, Andrew J. Kauffman, and William L. Putnam. *A Century of American Alpinism.* Boulder, Colo.: American Alpine Club, 2002.

Fichman, Martin. *An Elusive Victorian: The Evolution of Alfred Russel Wallace.* Chicago: University of Chicago Press, 2004.

Finnegan, Diarmid A. "Natural History Societies in Late Victorian Scotland and the Pursuit of Local Civic Science." *British Journal for the History of Science* 38 (2005): 53–72.

———. "The Spatial Turn: Geographical Approaches in the History of Science." *Journal of the History of Biology* 41 (2008): 369–88.

Frenkel, Stephen. "Geography, Empire, and Environmental Determinism." *Geographical Review* 82, no. 2 (1992): 143–53.

Galison, Peter. "Judgment against Objectivity." In *Picturing Science, Producing Art*, ed. Caroline A. Jones and Peter Galison, 327–59. New York: Routledge, 1998.

———. "Trading Zone: Coordinating Action and Belief." In *The Science Studies Reader*, ed. Mario Biagioli, 137–60. New York: Routledge, 1999.

Gieryn, Thomas F. *Cultural Boundaries of Science: Credibility on the Line.* Chicago: University of Chicago Press, 1999.

Godlewska, Anne. "Map, Text and Image: The Mentality of Enlightened Conquerors: A New Look at the *Description De L'Egypte*." *Transactions of the Institute of British Geographers* 20 (1995): 5–28.

Godlewska, Anne, and Neil Smith. *Geography and Empire.* Oxford: Blackwell, 1994.

Goetzmann, William H. *Exploration and Empire: The Explorer and the Scientist in the Winning of the American West*. New York: Alfred A. Knopf, 1966.

Golinski, Jan. *Making Natural Knowledge: Constructivism and the History of Science*. Cambridge: Cambridge University Press, 1998.

Gooday, Graeme. "The Premisses of Premises: Spatial Issues in the Historical Construction of Laboratory Credibility." In *Making Space for Science: Territorial Themes in the Shaping of Knowledge*, ed. Crosbie Smith and Jon Agar, 216–45. New York: St. Martin's Press, 1998.

Gregory, Derek. "Between the Book and the Lamp: Imaginative Geographies of Egypt, 1849–50." *Transactions of the Institute of British Geographers* 20 (1995): 29–57.

———. "(Post)Colonialism and the Production of Nature." In *Social Nature: Theory, Practice and Politics*, ed. Noel Castree and Bruce Braun, 84–111. Oxford: Blackwell, 2001.

Grove, A. T., and Oliver Rackham. *The Nature of Mediterranean Europe: An Ecological History*. New Haven: Yale University Press, 2001.

Grove, Richard H. *Ecology, Climate, and Empire: Colonialism and Global Environmental History, 1400–1940*. Cambridge: White Horse Press, 1997.

———. "The Evolution of the Colonial Discourse on Deforestation and Climate Change, 1500–1940." In *Ecology, Climate and Empire*, 5–36. Cambridge: White Horse Press, 1997.

———. *Green Imperialism: Colonial Expansion, Tropical Island Edens and the Origins of Environmentalism, 1600–1860*: White Horse Press, 1995.

———. "Imperialism and the Discourse of Desiccation: The Institutionalisation of Global Environmental Concerns and the Role of the Royal Geographic Society, 1860–1880." In *Geography and Imperialism, 1820–1940*, ed. Morag Bell, Robin Butlin, and Michael Heffernan, 36–52. Manchester: Manchester University Press, 1995.

Guelke, Jeanne K., and Karen M. Morin. "Gender, Nature, Empire: Women Naturalists in Nineteenth-Century British Travel Literature." *Transactions of the Institute of British Geographers* 26 (2001): 306–26.

Guelke, Leonard, and Jeanne Kay Guelke. "Imperial Eyes on South Africa: Reassessing Travel Narratives." *Journal of Historical Geography* 30 (2004): 11–31.

Guthke, Karl S. *The Last Frontier: Imagining Other Worlds, from the Copernican Revolution to Modern Science Fiction*. Translated by Helen Atkins. Ithaca, N.Y.: Cornell University Press, 1983.

Hall, Stuart. "The West and the Rest: Discourse and Power." In *Formations of Modernity*, ed. Stuart Hall and Bram Gieben, 275–320. Oxford: Polity in Association with the Open University, 1992.

Hall, Stuart, and Bram Gieben, eds. *Formations of Modernity*. Oxford: Polity in Association with Open University, 1992.

Haraway, Donna J. *Modest_Witness@Second_Millennium.Femaleman©_Meets_Oncomouse™: Feminism and Technoscience*. New York: Routledge, 1997.

———. "Teddy Bear Patriarchy: Taxidermy in the Garden of Eden, New York City, 1908–1936." In *The Haraway Reader*, 151–98. New York: Routledge, 2004.

Harley, J. B. "Deconstructing the Map." *Cartographica* 26 (1989): 1–20.

———. "Maps, Knowledge, and Power." In *The Iconography of Landscape: Essays on the Symbolic Representation, Design and Use of Past Environments*, ed. Denis Cosgrove and Stephen Daniels, 277–312. Cambridge: Cambridge University Press, 1988.

———. *The New Nature of Maps: Essays in the History of Cartography*. Edited by Paul Laxton. Baltimore: Johns Hopkins University Press, 2001.

Hart, Roger. "Translating the Untranslatable: From Copula to Incommensurable Worlds." In *Tokens of Exchange: The Problem of Translation in Global Circulations*, ed. Lydia H. Liu, 45–73. Durham, N.C.: Duke University Press, 1999.

Heffernan, Michael J. "Bringing the Desert to Bloom: French Ambitions in the Sahara Desert During the Late Nineteenth Century—The Strange Case of 'La Mer Intérieure'." In *Water, Engineering and Landscape: Water Control and Landscape Transformation in the Modern Period*, ed. Denis Cosgrove and Geoff Petts, 94–114. London: Belhaven Press, 1990.

———. "The Limits of Utopia: Henri Duveyrier and the Exploration of the Sahara in the Nineteenth Century." *Geographical Journal* 155, no. 3 (1989): 342–52.

Heffernan, William C. "The Singularity of Our Inhabited World: William Whewell and A. R. Wallace in Dissent." *Journal of the History of Ideas* 39 (1978): 81–100.

Hegglund, Jon. "Empire's Second Take: Projecting America in *Stanley and Livingstone*." In *Nineteenth-Century Geographies: The Transformation of Space from the Victorian Age to the American Century*, ed. Helena Michie and Ronald R. Thomas, 265–77. New Brunswick, N.J.: Rutgers University Press, 2003.

Henson, Louise, Geoffrey Cantor, Gowan Dawson, Richard Noakes, Sally Shuttleworth, and Jonathan R. Topham, eds. *Culture and Science in the Nineteenth-Century Media*. Aldershot: Ashgate, 2004.

Hetherington, Norriss S. "Amateur Versus Professional: The British Astronomical Association and the Controversy over Canals on Mars." *Journal of the British Astronomical Association* 86 (1976): 303–8.

———. "Planetary Fantasies: Mars." In *Science and Objectivity: Episodes in the History of Astronomy*, 49–64. Ames: Iowa State University Press, 1988.

Hevly, Bruce. "The Heroic Science of Glacier Motion." *Osiris* 11 (1996): 66–86.

Hilgartner, Stephen. "The Dominant View of Popularization: Conceptual Problems, Political Uses." *Social Studies of Science* 20 (1990): 519–39.

Hill, Jen. *White Horizon: The Arctic in the Nineteenth-Century British Imagination*. Albany: State University of New York Press, 2008.

Hoyt, William Graves. *Lowell and Mars*. Tucson: University of Arizona Press, 1976.

Hudson, Brian. "The New Geography and the New Imperialism: 1870–1918." *Antipode* 9, no. 2 (1977): 12–19.

John, Gareth E. "Benevolent Imperialism: George Catlin and the Practice of Jeffersonian Geography." *Journal of Historical Geography* 30 (2004): 597–617.

———. "Cultural Nationalism, Westward Expansion and the Production of Imperial Landscape: George Catlin's Native American West." *Ecumene* 8, no. 2 (2001): 175–203.

Jones, Barrie W. "Mars before the Space Age." *International Journal of Astrobiology* 7, no. 2 (2008): 143–55.

Jones, Bessie Zaban, and Lyle Gifford Boyd. *The Harvard College Observatory: The First Four Directorships, 1839–1919*. Cambridge: Harvard University Press, 1971.

———. "Some Lofty Mountain." In *Harvard College Observatory: The First Four Directorships, 1839–1919*. Cambridge: Harvard University Press, 1971.

Kearns, Gerry. "The Imperial Subject: Geography and Travel in the Work of Mary Kingsley and Halford Mackinder." *Transactions of the Institute of British Geographers* 22, no. 4 (1997): 450–72.

Keller, Evelyn Fox. *Reflections on Gender and Science*. New Haven: Yale University Press, 1985.

Kennedy, Barbara A. "Inventing the Geographical Cycle and the Synthetic Genius of W. M. Davis." In *Inventing the Earth: Ideas on Landscape Development since 1740*, 87–97. Malden, Mass.: Blackwell, 2006.

Kennedy, Dane. "Imperial History and Post-Colonial Theory." *Journal of Imperial and Commonwealth History* 24, no. 3 (1996): 345–63.

Kenny, Judith T., ed. "Colonial Geographies: Accommodation and Resistance." Special issue, *Historical Geography* 27 (1999).

King, Anthony D. "(Post)Colonial Geographies: Material and Symbolic." *Historical Geography* 27 (1999): 99–118.

Kirsch, Scott. "John Wesley Powell and the Mapping of the Colorado Plateau, 1869–1879: Survey Science, Geographical Solutions, and the Economy of Environmental Values." *Annals of the Association of American Geographers* 92 (2002): 548–72.

Kitchin, Rob, and James Neale. "Science Fiction or Future Fact? Exploring Imaginative Geographies of the New Millennium." *Progress in Human Geography* 25, no. 1 (2001): 19–35.

Kohler, Robert E. *Landscapes and Labscapes: Exploring the Lab-Field Border in Biology.* Chicago: University of Chicago Press, 2002.

Kuklick, Henrika, and Robert E. Kohler, eds. "Science in the Field." Special issue, *Osiris*, 2nd ser., 11 (1996).

Kurashige, Lon. "Immigration, Race, and the Progressives." In *A Companion to California History*, ed. William Deverell and David Igler, 278–91. Oxford: Blackwell, 2008.

Lambert, David, Luciana Martins, and Miles Ogborn, eds. "Historical Geographies of the Sea." Special issue, *Journal of Historical Geography* 32 (2006).

Lane, K. Maria D. "Astronomers at Altitude: Mountain Geography and the Cultivation of Scientific Legitimacy." In *High Places: Cultural Geographies of Mountains, Ice, and Science*, ed. Denis Cosgrove and Veronica della Dora. London: IB Tauris, 2008.

———. "Geographers of Mars: Cartographic Inscription and Exploration Narrative in Late Victorian Representations of the Red Planet." *Isis* 96 (2005): 477–506.

———. "Mapping the Mars Canal Mania: Cartographic Projection and the Creation of a Popular Icon." *Imago Mundi* 58 (2006): 198–211.

Lankford, John. "Amateurs and Astrophysics: A Neglected Aspect in the Development of a Scientific Specialty." *Social Studies of Science* 11 (1981): 275–303.

———. "Amateurs Versus Professionals: The Controversy over Telescope Size in Late Victorian Science." *Isis* 72 (1981): 11–28.

———. *American Astronomy: Community, Careers, and Power, 1859–1940.* Chicago: University of Chicago Press, 1997.

Latour, Bruno. *Science in Action: How to Follow Scientists and Engineers through Society.* Milton Keynes: Open University Press, 1987.

———. *We Have Never Been Modern.* Translated by Catherine Porter. New York: Harvester Wheatsheaf, 1993.

Latour, Bruno, and Steve Woolgar. *Laboratory Life: The Construction of Scientific Facts.* Translated by Jonas Salk. 2nd ed. Princeton, N.J.: Princeton University Press, 1986.

Lee, Lawrence B. "William Ellsworth Smythe and the Irrigation Movement: A Reconsideration." *Pacific Historical Review* 41, no. 3 (1972): 289–311.

Lightman, Bernard. "The Visual Theology of Victorian Popularizers of Science: From Reverent Eye to Chemical Retina." *Isis* 91 (2000): 651–80.

———. "'The Voices of Nature': Popularizing Victorian Science." In *Victorian Science in Context*, ed. Bernard Lightman, 187–211. Chicago: University of Chicago Press, 1997.

Lipset, Seymour Martin. *American Exceptionalism: A Double-Edged Sword.* New York: W. W. Norton, 1997.

Livingstone, David N. "Climate's Moral Economy: Science, Race and Place in Post-Darwinian British and American Geography." In *Geography and Empire*, ed. Anne Godlewska and Neil Smith, 132–54. Oxford: Blackwell, 1994.

———. "Environment and Inheritance: Nathaniel Southgate Shaler and the American Frontier." In *The Origins of Academic Geography in the United States*, ed. Brian Blouet, 123–38. Hamden: Archon Books, 1981.

———. "The Geographical Experiment: Evolution and Founding of a Discipline." In *The Geographical Tradition*, 177–215. Oxford: Blackwell, 1993.

———. *The Geographical Tradition*. Oxford: Blackwell, 1993.

———. "Geography, Tradition, and the Scientific Revolution: An Interpretative Essay." *Transactions of the Institute of British Geographers* 15 (1990): 359–73.

———. *Putting Science in Its Place: Geographies of Scientific Knowledge*. Chicago: University of Chicago Press, 2003.

———. "Science, Text, and Space: Thoughts on the Geography of Reading." *Transactions of the Institute of British Geographers* 30 (2005): 391–401.

———. "The Spaces of Knowledge: Contributions Towards a Historical Geography of Science." *Environment and Planning D: Society and Space* 13 (1995): 5–34.

———. "A 'Sternly Practical Pursuit': Geography, Race and Empire." In *The Geographical Tradition*, 216–59. Oxford: Blackwell, 1993.

Lowe, Lisa. *Critical Terrains: French and British Orientalisms*. Ithaca, N.Y.: Cornell University Press, 1991.

Lutz, Catherine A., and Jane Collins. *Reading National Geographic*. Chicago: University of Chicago Press, 1993.

MacDonald, Fraser. "Anti-*Astropolitik*—Outer Space and the Orbit of Geography." *Progress in Human Geography* 31, no. 5 (2007): 592–615.

Macfarlane, Robert. *Mountains of the Mind*. New York: Pantheon Books, 2003.

MacKenzie, John M. *Imperialism and the Natural World*. Manchester: Manchester University Press, 1990.

Markley, Robert. *Dying Planet: Mars in Science and the Imagination*. Durham, N.C.: Duke University Press, 2005.

Marsh, George P. *Man and Nature, or, Physical Geopgrahy as Modified by Human Action*. New York: Charles Scribner 1864

Mayhew, Robert J. "Historical Geography 2007–2008: Foucault's Avatars—Still in (the) Driver's Seat." *Progress in Human Geography* 33, no. 3 (2009): 387–97.

McCook, Stuart. "'It May Be Truth, But It Is Not Evidence': Paul Du Chaillu and the Legitimation of Evidence in the Field Sciences." In "Science in the Field," ed. Henricka Kuklick and Robert E. Kohler, 177–97. Special issue, *Osiris*, 2nd ser., 11 (1996).

McEwan, Cheryl. *Geography, Gender, and Empire: Victorian Women Travellers in West Africa*. Aldershot: Ashgate, 2000.

McKim, Richard. "Nathaniel Everett Green: Artist and Astronomer." *Journal of the British Astronomical Association* 114, no. 1 (2004): 13–23.

———. "The Life and Times of E. M. Antoniadi, 1870–1944. Part 1: An Astronomer in the Making." *Journal of the British Astronomical Association* 103, no. 4 (1993): 164–70.

———. "The Life and Times of E. M. Antoniadi, 1870–1944. Part 2: The Meudon Years." *Journal of the British Astronomical Association* 103, no. 5 (1993): 219–27.

Mills, S. *Discourses of Difference: An Analysis of Women's Travel Writing and Colonialism*. London: Routledge, 1991.

Monmonier, Mark S. *Rhumb Lines and Map Wars: A Social History of the Mercator Projection*. Chicago: University of Chicago Press, 2004.

Morin, Karen M. "British Women Travellers and Construction of Racial Difference across the Nineteenth-Century American West." *Transactions of the Institute of British Geographers* n.s. 23 (1998): 311–30.

———. "Charles P. Daly's Gendered Geography, 1860–1890." *Annals of the Association of American Geographers* 98, no. 4 (2008): 897–919.

———. "Peak Practices: Englishwomen's 'Heroic' Adventures in the Nineteenth-Century American West." *Annals of the Association of American Geographers* 89, no. 3 (1999): 489–514.

———. "Trains through the Plains: The Great Plains Landscape of Victorian Women Travelers." *Great Plains Quarterly* 18 (1998): 235–56.

Morin, Karen M., Robyn Longhurst, and Lynda Johnston. "(Troubling) Spaces of Mountains and Men: New Zealand's Mount Cook and Hermitage Lodge." *Social and Cultural Geography* 2 (2001): 117–39.

Nash, Roderick. *Wilderness and the American Mind.* 3rd ed. New Haven: Yale University Press, 1982.

Naylor, Simon. "Historical Geography: Geographies and Historiographies." *Progress in Human Geography* 32 (2008): 254–82.

———. "Historical Geography: Knowledge, in Place and on the Move." *Progress in Human Geography* 29, no. 5 (2005): 626–34.

———. "Introduction: Historical Geographies of Science—Places, Contexts, Cartographies." *British Journal for the History of Science* 38 (2005): 1–12.

Nisbett, Catherine Elaine. "Business Practice: The Rise of American Astrophysics, 1859–1919." PhD diss., Princeton University, 2007.

Ophir, Ada, and Steven Shapin. "The Place of Knowledge: A Methodological Survey." *Science in Context* 4 (1991): 3–21.

Orde, Anne. *The Eclipse of Great Britain: The United States and British Imperial Decline, 1895–1956.* New York: St. Martin's Press, 1996.

Osterbruck, Donald E. "To Climb the Highest Mountain: W. W. Campbell's 1909 Mars Expedition to Mount Whitney." *Journal for the History of Astronomy* 20 (1989): 77–97.

———. *Yerkes Observatory, 1892–1950.* Chicago: University of Chicago Press, 1997.

Osterbruck, Donald E., John R. Gustafson, and W. J. Shiloh Unruh. *Eye on the Sky: Lick Observatory's First Century.* Berkeley: University of California Press, 1988.

Outram, Dorinda. "New Spaces in Natural History." In *Cultures of Natural History,* ed. N. Jardine, J. A. Secord and E. C. Spary, 249–65. Cambridge: Cambridge University Press, 1996.

Pang, Alex Soojung-Kim. "Gender, Culture, and Astrophysical Fieldwork: Elizabeth Campbell and the Lick Observatory-Crocker Eclipse Expeditions." In "Science in the Field," ed. Henricka Kuklick and Robert E. Kohler, 17–43. Special issue, *Osiris,* 2nd ser., 11 (1996).

———. "The Social Event of the Season: Solar Eclipse Expeditions and Victorian Culture." *Isis* 84, no. 2 (1993): 252–77.

Pedynowski, Dena. "Science(s)—Which, When and Whose? Probing the Metanarrative of Scientific Knowledge in the Social Construction of Nature." *Progress in Human Geography* 27 (2003): 735–52.

Peet, Richard. "The Social Origins of Environmental Determinism." *Annals of the Association of American Geographers* 75, no. 3 (1985): 309–33.

Pfitzer, Gregory M. "The Only Good Alien Is a Dead Alien: Science Fiction and the Metaphysics of Indian-Hating on the High Frontier." *Journal of American Culture* 18, no. 1 (1995): 51–67.

Pilkington, Margaret. "The Ecologist's Very Own Ecotone: Exploring the Lab-Field Border." *Journal of Biogeography* 31 (2004): 516.

Plotkin, Howard. "Harvard College Observatory's Boyden Station in Peru: Origin and Formative Years, 1879–1898." In *Mundialización de la Ciencia Y Cultura Nacional. Actas del Congreso Internacional 'Ciencia, Descubrimiento y Mundo Colonial,'* ed. A. Lafuente, A. Elena, and M. L. Ortega. Madrid: Doce Calles, 1991.

Powell, Richard C. "Geographies of Science: Histories, Localities, Practices, Futures." *Progress in Human Geography* 31, no. 3 (2007): 309–29.

Prakash, Gyan. *Another Reason: Science and the Imagination of Modern India*. Princeton, N.J.: Princeton University Press, 1999.

Pratt, Mary Louise. *Imperial Eyes: Travel Writing and Transculturation*. London: Routledge, 1992.

Rose, Gillian. "Geography as the Science of Observation: The Landscape, the Gaze, and Masculinity." In *Nature and Science: Essays in the History of Geographical Knowledge*, ed. Felix Driver and Gillian Rose, 8–18, 1992.

Rothenberg, Marc. "Organization and Control: Professionals and Amateurs in American Astronomy, 1899–1918." *Social Studies of Science* 11 (1981): 305–25.

Ryan, James R. "'Our Home on the Ocean': Lady Brassey and the Voyages of the Sunbeam, 1974–1887." *Journal of Historical Geography* 32 (2006): 579–604.

Ryan, Simon. *The Cartographic Eye: How Explorers Saw Australia*. Cambridge: Cambridge University Press, 1996.

———. "Inscribing the Emptiness: Cartography, Exploration, and the Construction of Australia." In *De-Scribing Empire: Post-Colonialism and Textuality*, ed. Chris Tiffin and Alan Lawson, 115–30. London: Routledge, 1994.

Saberwal, Vasant K. "Science and the Desiccationist Discourse of the 20th Century." *Environment and History* 3 (1997): 309–43.

Said, Edward W. *Culture and Imperialism*. New York: Vintage Books, 1993.

———. *Orientalism*. New York: Pantheon Books, 1978.

Schaffer, Simon. "Astronomers Mark Time: Discipline and the Personal Equation." *Science in Context* 2 (1988): 115–45.

———. "Physics Laboratories and the Victorian Country House." In *Making Space for Science: Territorial Themes in the Shaping of Knowledge*, ed. Crosbie Smith and Jon Agar, 149–80. New York: St. Martin's Press, 1998.

Schama, Simon. *Landscape and Memory*. New York: Alfred A. Knopf, 1995.

Schroeder, David. "A Message from Mars: Astronomy and Late-Victorian Culture." PhD diss., Indiana University, 2002.

Schulten, Susan. *The Geographical Imagination in America, 1880–1950*. Chicago: University of Chicago Press, 2001.

Schwartz, Joan M. "*The Geography Lesson*: Photographs and the Construction of Imaginative Geographies." *Journal of Historical Geography* 22, no. 1 (1996): 16–45.

Secord, Anne. "Science in the Pub: Artisan Botanists in Early Nineteenth-Century Lancashire." *History of Science* 32 (1994): 269–315.

Secord, James A. *Victorian Sensation: The Extraordinary Publication, Reception, and Secret Authorship of Vestiges of the Natural History of Creation*. Chicago: University of Chicago Press, 2000.

Shapin, Steven. "The House of Experiment in Seventeenth-Century England." *Isis* 79 (1988): 373–404.

———. "Placing the View from Nowhere: Historical and Sociological Problems in the Location of Science." *Transactions of the Institute of British Geographers* 23 (1998): 5–12.

Shapin, Steven, and Simon Schaffer. *Leviathan and the Air-Pump: Hobbes, Boyle, and the Experimental Life*. Princeton, N.J.: Princeton University Press, 1985.

Sheehan, William. *The Planet Mars: A History of Observation and Discovery*. Tucson: University of Arizona Press, 1996.

———. *Planets and Perception*. Tucson: University of Arizona Press, 1988.

Shermer, Michael. *In Darwin's Shadow: The Life and Science of Alfred Russel Wallace*. Oxford: Oxford University Press, 2002.

Shinn, T., and R. Whitley, eds. *Expository Science: Forms and Functions of Popularisation.* Sociology of the Sciences Yearbook. Dordrecht: Reidel, 1985.

Sinnott, Roger. "Mars Mania of Oppositions Past." *Sky and Telescope* (September 1988): 244–46.

Smith, Crosbie, and Jon Agar. *Making Space for Science: Territorial Themes in the Shaping of Knowledge.* New York: St. Martin's Press, 1998.

Smith, Neil. *American Empire: Roosevelt's Geographer and the Prelude to Globalization.* Berkeley: University of California Press, 2003.

———. "The Lost Geography of the American Century." In *American Empire: Roosevelt's Geographer and the Prelude to Globalization,* 1–30. Berkeley: University of California Press, 2003.

Stern, Alexandra Minna. *Eugenic Nation: Faults and Frontiers of Better Breeding in Modern America.* Berkeley: University of California Press, 2005.

Stoddart, D. R. "Darwin's Impact on Geography." *Annals of the Association of American Geographers* 56, no. 4 (1966): 683–98.

Strauss, David. "'Fireflies Flashing in Unison': Percival Lowell, Edward Morse, and the Birth of Planetology." *Journal for the History of Astronomy* 24 (1993): 157–69.

———. *Percival Lowell: The Culture and Science of a Boston Brahmin.* Cambridge: Harvard University Press, 2001.

———. "Percival Lowell, W. H. Pickering and the Founding of the Lowell Observatory." *Annals of Science* 51 (1994): 37–58.

Stromquist, Shelton. *Re-Inventing "The People": The Progressive Movement, the Class Problem, and the Origins of Modern Liberalism.* Urbana: University of Illinois Press, 2006.

Taylor, Peter J., Michael Hoyler, and David M. Evans. "A Geohistorical Study of 'the Rise of Modern Science': Mapping Scientific Practice through Urban Networks, 1500–1900." *Minerva* 46 (2008): 391–410.

Thomas, William L., Jr., ed. *Man's Role in Changing the Face of the Earth.* Chicago: University of Chicago Press, 1956.

Thrift, Nigel, Felix Driver, and David Livingstone. "The Geography of Truth." *Environment and Planning D: Society and Space* 13 (1995): 1–3.

Topham, Jonathan R. "A View from the Industrial Age." *Isis* 95 (2004): 431–42.

Tuason, Julie A. "The Ideology of Empire in National Geographic Magazine's Coverage of the Philippines, 1898–1908." *Geographical Review* 89, no. 1 (1999): 34–53.

Tucker, Jennifer. *Nature Exposed: Photography as Eyewitness in Victorian Science.* Baltimore: Johns Hopkins University Press, 2005.

———. "Photographic Evidence and Mass Culture." In *Nature Exposed: Photography as Eyewitness in Victorian Science,* 194–233. Baltimore: Johns Hopkins University Press, 2005.

———. "Voyages of Discovery on Oceans of Air: Scientific Observation and the Image of Science in an Age of 'Balloonacy'." In "Science in the Field," ed. Henricka Kuklick and Robert E. Kohler, 144–76. Special issue, *Osiris,* 2nd ser., 11 (1996).

Westcoat, James L., Jr. "Wittfogel East and West: Changing Perspectives on Water Development in South Asia and the US, 1670–2000." In *Cultural Encounters with the Environment: Enduring and Evolving Geographic Themes,* ed. Alexander B. Murphy and Douglas L. Johnston, 109–32. Lanham, Md.: Rowman and Littlefield, 2000.

Whitley, R. "Knowledge Producers and Knowledge Acquirers: Popularisation as a Relation between Scientific Facts and Their Publics." In *Expository Science: Forms and Functions of Popularisation, Sociology of the Sciences Yearbook,* ed. T. Shinn and R. Whitley, 3–28. Dordrecht: Reidel, 1985.

Winlow, Heather. "Mapping Moral Geographies: W. Z. Ripley's Races of Europe and the United States." *Annals of the Association of American Geographers* 96, no. 1 (2006): 119–41.

Withers, Charles W. J. "History and Philosophy of Geography, 2002–2003: Geography in Its Place." *Progress in Human Geography* 29 (2005): 63–72.

Wittfogel, Karl A. "The Hydraulic Civilizations." In *Man's Role in Changing the Face of the Earth*, ed. William L. Thomas Jr., 152–64. Chicago: University of Chicago Press, 1956.

———. *Oriental Despotism: A Comparative Study of Total Power*. New Haven: Yale University Press, 1957.

Wood, Denis, and John Fels. "Designs on Signs: Myths and Meaning in Maps." *Cartographica* 23, no. 3 (1986): 54–103.

———. *The Natures of Maps: Cartographic Constructions of the Natural World*. Chicago: University of Chicago Press, 2008.

Worster, Donald. *Rivers of Empire: Water, Aridity, and the Growth of the American West*. New York: Pantheon Books, 1985.

———. *A River Running West: The Life of John Wesley Powell*. Oxford: Oxford University Press, 2002.

Youngs, Tim. "'My Footsteps on These Pages': The Inscription of Self and 'Race' in H. M. Stanley's *How I Found Livingstone*." *Prose Studies* 13 (1990): 230–49.

INDEX